The Herb Manual

© 2015 GoMidwife. All rights reserved.

ISBN 978-1-329-62893-9

Introduction	4
Alternative Medicine	5
Philosophy	6
The Garden	7
Experts in Health	9
Heal Yourself	10
Symptoms vs. Cause	12
Holistic Approach	13
The Disease Based Model	14
Introduction to Herbs	16
Do Herbs Work	17
Are Herbs Safe	18
Why Herbs Work	19
Herbology	20
Functions of Herbs	22
Herbal Vocabulary	22
How to Use Herbs	27
Culinary and Medicinal Herbs	31
Nourishment	32
Nourishing Herbs	32
Herbs for the Systems	47
Circulatory	47
Nervous	47
Lymphatic	48
Respiratory	48

- Immune 48
- Liver and Skin 52
- Urinary Tract 52
- Muscular and Skeletal 53
- Digestion 53

Herbal First Aid 62

Herbs for Children 69

Herbs for Sleep and Anxiety 79

Herbs for Women 82

Herbs for Reproduction 90

Herbs for Pregnancy 98

Herbs for Menopause 98

Herbs for Labor 102

Herbs for Postpartum 104

Herbs for Breastfeeding 108

Herbs for Newborn 112

Herbs de Toilette 117

Herbs for the Household 123

Foraging Herbs 125

More Recipes From My Herb Cabinet 126

> *"The doctor of the future will give no medicine, but will interest his patients in the care of the human frame, in diet, and in the cause and prevention of disease."*
> — Thomas Edison

Introduction

Alternative medicine and therapies cover a broad range of philosophies and approaches including chiropractics, homeopathy, acupuncture and pressure among others, and is generally defined as treatments and practices outside of traditional Western medicine. This includes herbal preventatives and restoratives, which is the approach we will look most closely at in this course.

> *There is no one who has more of a vested interest in keeping you well, and keeping your children well, than you. There is no doctor who will work harder to understand cause and concern and pursue a long-term solution to your health, than you.*
> — Amy Kirbow, CPM

I began to study and explore herbs back in 1997 after a trip to England revolutionized my understanding of health, well-being and not only the ability to care for myself, but the necessity. It was there I began to learn no one is more invested in me or my family than I am, most certainly not a doctor whom I spend at most twenty minutes with prior to the doling out of this pill or that. Doctors are not looking for a solution, or an underlying cause, doctors generally treat and suppress symptoms, with no understanding, nor desire to understand underlying cause. It was a revolutionary idea that I could take control of my health and with careful study, could understand most underlying causes to symptoms and choose a natural remedy with which to begin a foundational treatment. Even more, was the understanding of prevention. Western medicine rarely, if ever, looks at prevention as the key to health. The foundation of herbal alternatives is prevention first, and treatment second, with the desire to plumb the depths and reach true understanding of cause and not merely medicating the symptoms alone.[i]

*Healing comes from nature and not from the physician.
Therefore the physician must start from nature with an open mind.*
Paracelsus

Alternative Medicine

Choosing physician care over holistic care is not a new problem. We see many scholars discussing this subject throughout the time-line of history. However, it is a problem which has perpetually become worse in the last several decades.

med·i·cine[ii]
ˈmedəsən/
noun

1.
    the science or practice of the diagnosis, treatment, and prevention of disease (in technical use often taken to exclude surgery).
2.
    a compound or preparation used for the treatment or prevention of disease, especially a drug or drugs taken by mouth.

Medicine as it is defined is not the problem, neither is medicine as it is intended, but medicine as it is practiced is on a precarious slope which we have now tipped.[iii] The unrestrained and tumultuous landslide we have caused shows no sign of slowing, only gaining momentum. It is the individual alone who bares the responsibility of saying no and in so doing, determines a different course.

There is an alternative to synthetic chemicals and that is nature and the bounty of health she provides. In fact, many of the synthetic pharmaceuticals are derived originally from the properties of herbs. They are selling back to us, that which God planted all around us.

"The art of medicine consists of amusing the patient while Nature cures the disease."

Voltaire

Philosophy

Prevention is not the aim of modern and contemporary medicine; it is however, the intention of herbal remedies. True health is predicated upon prevention. Unlike conventional medicine, herbal medicine does not ignore the symptoms, but rather seeks to discover and eliminate the elemental cause while also addressing the symptoms. Herbs are tonic in nature, rectifying systemic mineral deficiencies, strengthening weakened body systems and providing the foundation needed for the body to heal itself.[iv] The approach of herbalism then is constructed on the understanding as follows: when all we know as modern medicine is stripped to its core we are left with plants and if that is where we would begin, then we would prevent more illnesses than would need treated[v].

It is essential to examine the natural resources around us and reacquire the wisdom and understanding once so common. Plants contain phytochemicals which hold immeasurable healing properties and they are everywhere; even the common weed lends itself to our well-being. We have strayed too far from the collective knowledge of medicinal plants, though every generation before us has known, we do not. In the age of dependency and convenience we have allowed ourselves to lose one of the earths most valuable gifts, and that is the understanding of the medicinal benefits of plant-life. When man allows for other men to decide for him, the result is never to the benefit of that man. Through choosing doctors and pharmaceutical companies to treat us we have isolated ourselves from the very resource and fountain of health which lies outside our door[vi]. The closer we live to nature as an individual, the healthier we are as a society. Nature is healing not only physically, but mentally, emotionally and spiritually. When we aspire to the pursuit wholeness we must reestablish our familiarity to plants and herbs[vii].

It is better to trust in the LORD than to put confidence in man.
Psalm 118:8

"And God said, Behold, I have given you every herb bearing seed, which is upon the face of all the earth, and every tree, in the which is the fruit of a tree yielding seed; to you it shall be for meat."
Genesis 1:29

The Garden

God created man and placed them in the _____ surround by all they needed to remain wholly _____ and in _____ to Him.

God has an affinity for gardens just as he has for birth. It was the two instructions he gave us: take care of the land and have babies.

> **God blessed them and said, "Be fertile and increase in number...**
> Genesis 1:28

> **The LORD God took the man and put him in the Garden of Eden to work it and take care of it.**
> Genesis 2:15

In the garden, in the cool of the evening, God walked and talked with man. The garden was the culmination of the best of what God, the Creator, had to offer his creation. Today demands a return, for a myriad of reasons, not least among them being our health: physical, mental and spiritual. All of our needs were met in the garden, every one: our food, medicine, exercise, water and a quiet place for communion with God. We must once again walk in nature with God and listen to what he says, we must ask him for the secrets to the abundant health he has planted around us. It was no accident man was originally placed in the garden and the further man has removed himself from the garden the greater increase in disease. The modern healthcare system is the antithesis of the intent of God in the beginning. It is warfare and our bodies, our minds and spirits are the battleground. We were created for the garden and we are called to return to garden relationship in order to restore the foundations of health.

© GoMidwife 2015

"At the end of times the merchants of the world will deceive the nations through their sorcery."
Revelations 18:23

This following subject has been met with some resistance and certainly a bit of controversy. However, I believe it to be quite true. We are spiritual beings and at the root of all is a spiritual conflict. If we are mind body and spirit, then why wouldn't our well-being, both mental and physical, be the battle ground? If our bodies and our minds can be corrupted with disease or numbed and neutralized through the consumption of chemicals then we are completely ineffective and unproductive in the work of the world and the kingdom.

Sorcery is translated pharmacia or pharmakia which is from the Greek and simply put means sorcery or magic spells disguised as medicine. Our modern words pharmacy and pharmaceutical are derived from the original Greek word and is directly translated as follows:

pharmakeia: the use of medicine, drugs or spell
Original Word: φαρμακεία, ας, ἡ
Part of Speech: Noun, Feminine
Transliteration: pharmakeia
Phonetic Spelling: (far-mak-i'-ah)
Short Definition: magic, sorcery, enchantment
Definition: magic, sorcery, enchantment.

Word Origin
from pharmakeuó (to administer drugs)
Definition
the use of medicine, drugs or spells[viii]

The more pharmaceuticals we consume, the more we are under the magic spell of the world.[ix] Pharmaceuticals manipulate our minds and binds up our bodies. Pharmaceuticals are in direct opposition to the original intent of the Lord in the garden, which was for us to take from every tree and every plant, walk with Him and be whole.

> "Most over-the-counter and almost all prescribed drug treatments merely mask symptoms or control health problems or in some way alter the way organs or systems such as the circulatory system work. Drugs almost never deal with the reasons why these problems exist, while they frequently create new health problems as side effects of their activities."
>
> John R. Lee, M.D

Experts in Health

No one is a better expert in your health than you. You know your body, what is normal, and what is not. We should not be so quick to give over that expertise to someone who does not know us holistically. There are times we need to seek out expert physicians, but it is not for every common cold. We may never be cardiologists, but we can understand how to maintain heart health, we may never be immunologists, but we can understand the immune system, how it is primed and how it is suppressed. Whereas doctors are all to often experts in disease, we can be experts in health. Far too often, when we seek expert advise the answer they provide is in a pill, where a better answer could be found in whole foods and herbs which provide the micro-nutrients from which our bodies are starved.

According to the Mayo Clinic:

Nearly 70 percent of Americans are on at least one prescription drug, and more than half take two. Those drugs include: antibiotics, antidepressants and painkilling opioids. Of the seventy percent who take daily medications **twenty percent are on five or more prescription medications**.[x]

We must determine to become experts in our own health, because no one else will. Their pervasive answer, which is a non-answer and only exacerbates the problem, is one more pill. Our eyes must be opened to see we have been swindled and the grand illusion of modern healthcare is in fact, what is making us sick.

"The next major advance in the health of the American people will be determined by what the individual is willing to do for himself."
John Knowle

Heal Yourself

The human body has a vast and unprecedented ability to _____ _____.

Modern healthcare is not the answer. Pharmaceuticals serve mainly to depress our systems in order to suppress our symptoms. This does not promote healing nor health. The individual must begin to take responsibility for their own care. Our bodies are designed to be well and when allowed the capacity to heal, will.

Within the next year, 98% of the cells in your body will die and be replaced. Each type of cell runs through this cycle at its own speed. The cells that make up your stomach lining, for instance, complete the cycle of renewal every 5 days. For your blood cells, the cycle takes 4 months. Even your brain renews itself—that cycle takes a full year.

Present society has created a dependency on others to provide answers for them, we expect a quick and easy fix to any uncomfortable situation. It is our experience that pain is bad, but not all pain is unproductive. When a hand is placed too close to a fire discomfort immediately tells us to stop that action, but when we feel pain or discomfort with an acute illness and immediately suppress it through medications, then the signals are turned off, our bodies no longer know how to respond to protect us, and illness becomes chronic. So often healthcare by its very nature is defined as something that is done to us, not something in which we participate. When we allow others to make decisions for us and think for us, we ourselves become passive in our own health and we pay the price through chronic illness and disease. When we choose not to participate we forfeit our original design and render if void. No longer can we be a silent audience to our own disease. We must begin to actively participate in our own health and well-being. Our bodies are designed to regenerate when given the proper care and fuel. Active participation means we must earnestly be involved in our health through diet, exercise and preventative measures not only for the health of our bodies, but also of our minds and spirits[xi].

> "The greatest discovery of any generation is that human beings can alter their lives by altering the attitudes of their minds."
> Albert Schweitzer

Notes

A bodily disease, which we look upon as whole and entire within itself, may, after all be but a symptom of some ailment in the spiritual past.
Nathaniel Hawthorne

Symptoms vs. Cause

We often address the complaints of the physical body by_____ alone. Rarely do we seek to define the _____ _____.

Herbs do not generally address the acute symptoms, rather aims at the prevention of chronic disease. Although symptoms can be addressed through herbal therapy the symptoms are not the enemy, but rather the alert system that something is not right in our bodies. Suppressing the symptoms, without addressing the cause only leads to more disease. Herbal therapies seek to restore the body and stimulate the capacity for healing. This approach is a slower and often misunderstood as not effective. It is always the goal to prevent further disease rather than treat the symptom.

It is widely thought for the underlying cause of most chronic illnesses to be inflammation at the root. Simply masking the symptoms, without understanding the cause will inevitably lead to the worsening of one's overall health long term. Unless it is an acute onset, we witness symptoms usually after something has been wrong in the body, without being addressed, for period of time. Once those symptoms appear, if they are simply suppressed, then we are doing ourselves a huge disservice. Once symptoms occur we must take action to find the root and eliminate the cause. Symptoms alone are rarely ever the cause and when we eliminate the symptoms and not the cause, then we perpetuate illness. Unfortunately, symptoms can often look very much like a cause and we stop short before truly becoming well. In order to identify root cause we have to begin by asking why is the symptom occurring. The ability to find the root cause requires us to know our bodies. It takes time and requires active participation for us to become attuned to our bodies and sensitive to the signal and response, but we cannot neglect whole to quickly fix a part.

"The part can never be well unless the whole is well."
Plato

Holistic Approach

HO·LIS·TIC: characterized by the _____ of the parts of something as intimately interconnected and explainable only by reference to the _____.

The modern Western healthcare system sees our bodies as a _____ _____ not a _____; they then treat the fragment not the whole. Our approach to wellness must be _____.

Holistic medicine generally approaches wellness by looking at the mind (and emotions), body and spirit. This approach views the person as a whole and suggests if one part is not well, then no parts can be. Although holistic medicine is considered to be an alternative it is closer to the original design, and the intent of God, than any conventional or modern medicine. The systems within our bodies are created to be interdependent of one another, and when we allow ourselves to be viewed, or treated, as a fragmentation we often alleviate only the symptoms and miss the root cause.

Not only does modern medicine splinter the whole of our bodies dividing the mind, body and spirit, but the fracture continues deeper when the treatment of our bodies are broken down into further fragmented sections in order to be treated. Very few doctors can treat even our physical bodies holistically. If we have heart trouble we go to a cardiovascular physician, stomach pains calls for a gastroenterologist, immune disease and we must see an immunologist. No one doctor then can, or will, relate stomach inflammation with overall immunodeficiency and so often miss the root cause in favor of the recognized symptoms. Specialists look only at their specialty and rarely beyond. This is detrimental to our overall health and true wellness will rarely take place until we begin to treat, and demand others, treat our bodies holistically.

One of the biggest tragedies of human civilization is the precedents of chemical therapy over nutrition. It's a substitution of artificial therapy over nature, of poisons over food, in which we are feeding people poisons trying to correct the reactions of starvation..

Dr. Royal Lee

The Disease Based Model

With the fragmentation of treatment, comes further breakdown within the modern healthcare system. When the part is viewed the whole is forgotten easily allowing for the physician to treat the disease and not the person. It is a model based, not on prevention, not on holistic wellness, but on symptomatic bandaging. A disease-based model is a broken one. It serves the system only when we remain ill.

We cannot deny that illness is a natural process of life. Nor can we deny ourselves the time, ability or capacity to _____ .

We cannot be _____ from God and be _____ .

Disease is a process of life, but the critical challenge is to support the proper function of the body, stimulate regeneration, promote healing and not allow for the progression of the disease. We are designed to be self-healing and created to be well, but we must renegotiate our position.

"A merry heart does good like a medicine, but a broken spirit dries the bones."

Proverbs 17:22

Our physical health is often the reflection of our spiritual or emotional well-being. Where modern medicine would segment our minds from our bodies and our spirits from both, the antithesis of the disease-based model would be the Garden model. It is understanding the further we are removed from the presence of God, and his creation, the less opportunity we have to be whole. Our bodies reflect what we are filled with and when our days are filled with busyness, processed foods, chemicals, endless noise or negative emotions such as forgiveness, bitterness or anger our bodies cannot be well and disease is promoted. It serves us to return to the garden relationship, where we can be whole mind, body and spirit.

The art of plant-based wellness is using the natural world around us to be well. Knowing the properties of multiple plants and using those properties to make our own vitamins and medicines, but even beyond that incorporating whole foods such as plants, flowers and weeds into our daily diets and helping them find their way through daily walks in nature to our plates. This model bases the foundation of our well-being on plants, as I believe, was the original plan in the garden.

Notes_____

> *"Before the establishment of universities in the eleventh and twelfth centuries, monasteries served as medical schools".*
>
> The University of Virginia

Introduction to Herbs

Definition of an Herb

1._____

_____

2._____

_____

Herbs are a never-ending wonderment. If we began today and studied for fifty years, we would not be able to plumb the depths of all the plants and herbs in our world, and the endless nourishment they provide. Since ancient times, long before the advent of modern medicine, herbs have been used to prevent disease, treat ailments, restore our mind and body and help maintain overall well-being. The Lascaux cave paintings in France, dating back some 17,300 years ago detail images of herbal use, and the Chinese have documented through written and oral traditions the use of herbal medicine for more than 3,000 years. Many cultures throughout the history of time record the use of herbs for medicinal purposes. These cultures include the Egyptians, the Babylonians, the Native Americans, India, Greece, Rome, Europe where during the Middle Ages the Church kept the herbal traditions alive eventually bringing the use of herbs to the New World and our forefathers and mothers. Herbs are not a passing whim, but rather founded on time-tested traditions.

Herbs are considered to be aromatic, culinary and ornamental, each soothing a part of the human whole: mind, body and spirit. They work in ways we are still discovering, providing benefits that affect every aspect of our lives. The effectiveness, ease of use, storied tradition, and accessibility are only some of the reasons that they have outlived other preventatives, medicines, and treatments. With such a rich history and powerful background it is not surprising that the popularity of using herbs in a myriad of ways still abounds. There are, nevertheless, questions surrounding herbs and their use. It is here that we will begin to examine their efficacy and safety.

"All that man needs for health and healing has been provided by God in nature, the challenge of science is to find it."
Philippus

Do Herbs Work?

What makes the difference between herbs being food and herbs being medicine

_____   _____   _____   _____.

This is a common question asked, and the answer quite simply, is yes herbs work. However, we must understand how they work. Herbs do not work like modern pharmacological prescriptions. When using herbs you must seek the underlying cause of the disease and seek to treat the disease at its root, not the fruit of the disease, which are the symptoms. Modern medicine is designed to suppress our symptoms. Our bodies are designed to heal themselves with the proper amount of rest and nourishment, our bodies are most often self healing. When we suppress our symptoms, we also suppress our body's ability to respond. If we do not allow ourselves to have a fever, or feel pain, but rather immediately suppress those warning signs then our bodies are confused and do not know to send out the defense mechanism to protect and heal. Instead, the senses are lulled to sleep and the cells and neural pathways lack the ability to communicate what the body needs for healing.

Herbs therefore, are not a quick fix. They will require multiple and continued doses over a longer period of time, and to that end are often mistaken as not working. However, herbs work differently to bring healing and that approach must be understood. They will not generally suppress your symptoms, but rather support your body to be able to fight off the illness. Herbs are supportive and building. They allow your body the necessary nutrition for the cells to function and communicate properly for regeneration.

Herbs are the friend of the physician.

Charlemagne we

Are Herbs Safe?

The answer to this question is both yes and no. Yes, herbs are safe when taken correctly and they are also be very dangerous when used incorrectly. Herbs are medicines. They are just natural God-made medicines, not man-made synthetic chemicals. However, many plants are toxic or when combined can become toxic. It is very important to educate yourself thoroughly before become your own doctor, but all doctors at one time or another had to begin their education process and journey. It is of significant importance we become willing to educate ourselves, and do the necessary research to know what out body needs to be well.

Most of our common pharmaceutical drugs are derived from plants originally. Chemists take specific plants and isolate specific properties within those plants and either then create a synthetic form or isolation and extract one property of the plant. Herbs then are generally considered safer than chemically derived drugs. They are not synthetic and they are not isolations. When the plant is left as a whole there are abilities for synergistic compounds to work. While the pharmaceutical field has determined that an active property within a plant is the reason for its benefit, the truth often is much more complicated. Using an herb can therefore work in a number of different ways as various compounds promote and benefit one another throughout the various systems of the body. Herbs are medicinal which is why chemists turn to plants time and again; they have powerful healing properties. This is why their side effects, as well as allergic reactions and death, can occur when used improperly. Herbs are, however, considered far safer than their chemically derived conventional counterparts. When used correctly, they promote health. How this happens is where our discussion now moves.

Notes_____

_____

_____

_____

_____

> *"Today, more than 95% of all chronic disease is caused by food choice, toxic food ingredients, nutritional deficiencies and lack of physical exercise."*
> — Mike Adams

Why Herbs Work

Herbs work because they are holistically nourishing. They are not a miraculous cure-all, but rather work at the cellular level to feed our bodies with vital micro-nutrients commonly missing from our modern diets. Below is is a list of common vitamins and minerals. Look up each of the following and determine what role they have in our overall well-being at the cellular level.

Fill in the action each vitamin and mineral in the space provided.

Vitamin and Mineral	Action
Vitamin A	
Vitamin B	
Vitamin C	
Vitamin D	
Vitamin E	
Vitamin K	
Potassium	
Magnesium	
Iron	
Manganese	
Calcium	
Copper	
Zinc	
Selenium	
Iodine	
Essential Amino Acids	

"Dare to know! Have the courage to use your own intelligence!"

Immanuael Kant

Herbology

Is a cultivated understanding of how God_____ and uses nature to our overall benefit. Plants and herbs are valuable resources at the fingertips of every woman in the world. It is our_____ as God's creation to know and understand His creation better...and to make it known.

Simply put, herbology is the study of herbs and how they improve our lives. Herbology is a science. Science as defined is a system of study which organizes and builds upon its knowledge through testable predictions. Herbology studies the use of plants and their parts: roots, stems, leaves, seeds and fruit, throughout the history of time, as medicinal preventions and remedies for disease.

Most herbs have not been been studied extensively here in the U.S. and for that reason have not been approved by the Food and Drug Administration, which is the regulatory body of pharmaceuticals within our country. This does not mean herbs are not without medicinal properties, but most known benefits would be considered anecdotal or evidence based. Since the following herbs discussed have not been studied or approved as medication we will view them as supplemental nourishment only.

There is no doubt our modern diets are extremely lacking in vital nutrients, and when we bring herbs back to the table, as did a generation previous, when we begin to explore herbs as answers to the chronic disease facing our generation, then we will see the role of health they assume in our lives. Because so much misinformation abounds and even what we know is limited compared to those that came before, it takes careful study and understanding to bring the concepts of using herbs back into the mainstream. We do this through something known as Herbology. Herbology is a science which continues to be discredited because it is a concept which is vastly misunderstood. While controversy swirls around specific herbs, their potency, effectiveness, and so forth, the truth is that when studied carefully and prepared accordingly there is little doubt that the use of herbs in health is beneficial and verifiable[xii].

"A wise man ought to realize that health is his most valuable possession."

Hippocrates

Notes_____
_____
_____
_____
_____
_____
_____
_____
_____
_____
_____
_____
_____
_____
_____
_____
_____
_____
_____
_____
_____
_____
_____
_____
_____
_____
_____
_____
_____

"A man may esteem himself happy when that which is his food is also his medicine."
Henry David Thoreau

Herbal Terms and Functions

When working with anyone, but especially pregnant women and children, it is of the utmost importance to understand the action of each herb. Below is a list of common herbal actions. An herbal action refers to the expected effect an herb will have in the body. Using the space provided below fill in the definition of each action and then research to find which herbs can be used to elicit the desired action.

Herbal Action	Definition	Herbs that Cause this Action
Abortifacient		
Adaptogen		
Allopathy		
Alterative		
Analgesic		
Antacid		
Anthelmintic		

© GoMidwife 2015

Anti-inflammatory		
Antibacterial		
Antibilious		
Anticatarrhal		
Antiemetic		
Antilithic		
Antimicrobial		
Antioxidant		
Antioxidant:		
Antirheumatic		
Antiseptic		
Antispasmodic		
Aphrodisiac		
Astringent		
Bitter		

Cardiac Tonic		
Carminative		
Cathartic		
Cholagogue		
Deep Immune Activator		
Demulcent		
Depurative		
Diaphoretic		
Diuretic		
Emetic		
Emmenagogue		
Expectorant		
Febrifuge		
Galactogogue		

Hepatic		
Hypnotic		
Laxative		
Liniment		
Lymphagogue		
Mucilage		
Nervine		
Phytoestrogens		
Rubefacient		
Sedative		
Sialagogue		
Stimulant		
Tonic		

Lemon balm causes the heart and the mind to become merry.

Seraphio

Notes

> *"Drink your tea slowly and reverently, as if it is the axis on which the world earth revolves – slowly, evenly, without rushing toward the future."*
>
> Thich Nhat Hanh

How to Use Herbs

If herbs are to be used for nourishment or medicinally then they should always be made from the freshest herbs possible. The best way to insure your herbs are fresh when you begin making medicinals is to grow and harvest your own, if that is not possible then consider foraging. If you can neither grow nor forage then you can buy in bulk from a reputable company. Here are the most common ways to use herbs medicinally:

Tissane

Teas found in stores typically use 1/7 the amount of herb used in home brews. These teas are not curative or preventative in nature rather enjoyed for flavor.

Teas

Herbal teas are simply the extraction of the phyochemicals into water. Medicinal teas are one of the oldest, easiest and cheapest forms of medicine available. Not only does the herbal tea provide medicinal benefits, one must also rest while sipping tea and it also provides hydration. Generally herbal teas should be made to order, but can be stored in the refrigerator up to 3 days. Dried herbs= 1 ounce per 1 pint of water. Fresh herbs= double the amount of dried. Standard does is 1-3 cups taken daily.

Infusion

Simple infusions are teas made, most commonly, from the leaves, flowers or berries of a plant. Infusions are made by bringing water to a boil and pouring over the herbs to infuse or extract the nourishing benefits from the plant. The plant is then steeped until cool or cold and can be consumed immediately or allowed to continue steeping for several hours and even overnight.

Decoctions

These teas are usually made with the stems, bark, seeds or roots of a plant. This calls for macerating the herbs and boiling in water. This extracts deeper properties from the herbs by simmering them until the water is reduced by half in an uncovered pot. Or simmered for the same length of time in a covered pot preventing evaporation. The herb is placed in a pan of water and brought to a boil for 5 to 10 minutes then allowed to simmered for about 30 minutes up to a few hours. The amount of water will be reduced and more water will need to be added back prior to consumption.

Tinctures & Vinegar Extracts

Tinctures are most commonly made with either alcohol known as an extract or apple cider vinegar. A rule of thumb is if you want to pass the tincture down to future generations or take the tincture sparingly, use alcohol. If it is a tincture you will be used daily, then you might want to make it with apple cider vinegar as the base. The medicinal benefits of the plants are extracted into the liquid, the foliage discarded and the liquid consumed.

Steams

Steams are used to open, cleanse and detox. Steams can be used for respiratory infections, colds and flues, for facials and for feminine or menstrual cleanses. Herbs are infused into boiling water and the the escaping heat helps to draw and increase circulation to the area, opens up the pores releasing toxins, breaking up mucous, increasing relaxation and promoting overall health.

Poultice

Poultices are made from macerated herbs and used externally. Commonly poultices are used to reduce inflammation, draw out splinters and toxins, reduce bleeding or prevent infections.

Poultices are also used to alleviate cramps. Fresh herbs are macerated, and spread out over the affected area. If spicy herbs are used then a cloth needs to be placed around the herb to prevent direct contact with skin. Some of the easiest poultices are known as spit poultice and can be applied directly after chewing the herb.

Creams/Salves

Creams are some of the easiest ways to use herbs and last for quite a long time. Creams are made with infused oils and beeswax. It is made thick enough to remain on the skin. Creams can be made as toiletries in lieu of store bought products full of stabilizers or medicinal first aid balms.

Syrups

Herbal syrups and honeys are some of the best ways for children to take herbal medicines. Syrups are made by using strong infusions or decoctions and adding honey, pure maple syrup or vegetable glycerin.

> "Just a spoonful of sugar helps the medicine go down
> In a most delightful way."
> Mary Poppins

Honeys

It cannot be said enough just how easy it is to make herbal medicinals. In fact, herbal honeys may make the top of the list as far as ease. The one note of importance here is to use dry herbs. Some recipes will allow for fresh herbs, but due to the rare possibility of botulism, it is best to avoid the possibility altogether and just go for completely dried herbs here. Simply choose an herb and submerse it in honey. The herbs will infuse into the honey and can be taken by the spoonful or stirred into herbal teas.

Oxymels

Oxymels are bitter sweet medicinals made from a combination of a vinegar and sweetener, and herbs. Perhaps one of the oldest ways to consume herbs medicinally and available long before the advent of alcohol. While a vinegar extract will not distill as deeply as alcohol, it will sufficiently extract vitamins and minerals from the plant structure.

Elixir

An herbal elixir is simply an herbal honey mixed with brandy. The addition of brandy will extend the shelf life of the herbal and will also be warming in the case of a cold or flu.

Bathes

An herbal bath is infused herbs steeped into a warm bath for the purpose of detoxification, relaxation, osmosis or healing.

Notes _____

_____
_____
_____
_____
_____
_____
_____
_____
_____
_____
_____
_____
_____
_____
_____
_____
_____
_____
_____

> "Until man duplicates a blade of grass, nature can laugh at his so-called scientific knowledge. Remedies from chemicals will never stand in favour compared with the products of nature, the living cell of the plant, the final result of the rays of the sun, the mother of all life."
> T. A. Edison

Culinary and Medicinal Herbs

Culinary herbs simply refers to herbs which are commonly used for cooking, and medicinal herbs are curative, health-giving and restorative. Culinary herbs and medicinal herbs are simply herbs we can eat daily, which have healing properties. Herbs and botanicals not only lend culinary diversity to our food, but also provide a wealth of medicinal properties along the way. Food choices should be intentional, and aimed at maintaining our well-being not just feeding our hunger. Opportunity is lost when we choose foods that provide no nutritional benefit. The time is now to redefine the possibilities and increase our understanding of what home health care is and can be. As we rediscover our ability to prevent disease by what we put on a plate and bring to the table. Plate-based wellness is one which includes culinary and medicinal herbs. Health, as it is known today comes from the Old English word hale, which means "wholeness, being whole".[xiii] We cannot expect to be whole if what we consume is not. When we choose whole foods and include nourishing herbs, then our plate becomes our pharmacy and disease is diminished[xiv].

The healing properties of herbs support our systems by aiding their natural function. Imagine an exotic meal where digestion was stimulated, your immune system boosted and your heart was strengthened all while enjoying the delicious delights on the plate before you. These benefits occur when we consume common culinary herbs such as garlic, mint, ginger and turmeric. Culinary herbs with medicinal properties can be consumed daily and they serve, not only to keep us well, but to increase the depth and richness of our palette. Food then does become our medicine and the need for chemical based manufactured compounds is reduced. We are effectively preventing many of the issues that might have otherwise crept up in our lives, dealing with problems we are currently facing, while at the same time enjoying better nutrition, health, flavor, and the satisfaction of knowing that we have a strong hand in how we feel and our general wellness.

"Let food be thy medicine and medicine be thy food"
Hippocrates

Nourishing Herbs

We must mutually participate in our own _____: spiritual, mental and physical and that requires_____and _____.

Nourishing herbs are the_____ herbs to use, rarely have _____ _____and can be taken in _____quantities for _____periods of time.

Nourishing herbs are consumed more as _____than as _____.

Nourishment is the building block on which we can fashion a long and vigorous life. Herbs that provide our bodies foundational nourishment through depositing vital phytochemicals, vitamins and minerals in our cells and systems when consumed hold the keys to our overall health. Our bodies are designed to heal themselves and with the proper nourishment they will do just that. Herbs that can be consumed daily, and in larger quantities are considered to be nourishing herbs. These herbs are building, supporting, and toning in nature and work slowly and over the expanse of time. Used as food, nourishing herbs can be used daily and for long period of times.

Any herb which contains and provides the primary vitamins and minerals needed are considered to be nourishing and healing herbs. The connection between these nutrient dense herbs and our health cannot be ignored, and wellness becomes more sustainable as we return to the culinary heritage of whole foods, which includes large quantities of edible and nourishing herbs. Consuming herbs daily assures our body systems receive the proper support required to function at optimum capacity. Herbs and botanicals which can be added to our culinary routine and consumed daily include: parsley, nettles, garlic, ginger, rosemary, thyme, basil, oregano, sage, catnip, mint, elderberry, hibiscus, lemon balm and turmeric.

Parsley

Parts Used: leaves and roots.

Systems Supported: kidneys, bladder, stomach, liver, gallbladder, cardiovascular and immune.

Parsley is a <u>diuretic</u> which increases the flow of urine from the body, and is also a <u>tonic</u>. A tonic is an herb that promotes the overall function of the body. It is specifically a biliary tonic. Bile, as defined by the National Library of Medicine[1], is a fluid that is made and released by the liver and stored in the gallbladder. Bile helps with digestion and the breakdown of fats into fatty acids. This common garnish contains many vital vitamins, including Vitamin C, B12, K, A, folate (B9) and iron. Parsley helps flush out excess fluid from the body, thus supporting kidney function. Parsley is key to digestion, reduces flatulence and decreases mucous production in the gut. Traditionally parsley accompanied dinner to help aid digestions, but now is commonly added just for garnish and is not consumed, although a healthy habit of eating your garnish could help you better digest your meal. Parsley supports blood vessels and cardiovascular systems and can be used topically for bruising. The volatile oils in parsley can suppress overacting immune systems to help alleviate allergies, auto-immune disorders and chronic fatigue. Parsley is high in iron so consuming it helps build iron levels and reduces anemia. The bacterial benefits of parsley is that it can help kill the bacteria that cause bad breath and baldness brought on by bacteria and is know to help to increase the antioxidants in the blood stream. Parsley has balancing benefits in women with hormonal problems including the ability to improve estrogen levels, restoring blood in the uterus, and reducing premenstrual syndrome, dry skin and depression. There are warnings to be mentioned surrounding the consumption of parsley. As with most herbs parsley tea should not be consumed by pregnant women in excessive amounts or without consulting a health and wellness practitioner. Parsley can act as a uterus stimulant which can increase the risk of miscarriage. Also, for anyone consuming parsley at medicinal levels it should be said, excessive intake of parsley can cause nausea and vomiting or diarrhea. Parsley tea is abundant in oxalic acid, so it should be avoided by people who suffer from inflammation of the kidney inflammation.

---

1 National Library of Medicine, "Bile". Last accessed from https://www.nlm.nih.gov/ on 10/01/2015.

Nettles

Parts Used: Leaves and Roots

Systems Supported: urinary, respiratory, skeletal[xv], integumentary[xvi], circulatory, lymphatic[xvii] and endocrine

Nettles are <u>diuretic</u> and <u>astringent</u>. Nettle leaf is among the most valuable overall nourishing herb. Due to its many nutrients, stinging nettle is traditionally used as a spring tonic. It is a slow-acting nutritive herb that gently cleanses the body of metabolic wastes. It is one of the safest alternatives, especially in the treatment of chronic disorders that require long-term treatment. It has a gentle, stimulating effect on the lymphatic system, enhancing the excretion of wastes through the kidneys. Nettle leaf is effective at reducing symptoms of the digestive tract ranging from acid reflux, excess gas, nausea, colitis and Celiac disease. Nettles balance and support the adrenal glands, helping to reduce adrenal fatigue so common today. Nettles is extremely high in iron and can be used to compliment and alleviate anemia and anemia related fatigue thus increasing overall energy. Stinging nettle is beneficial during pregnancy due to its rich mineral value and vitamin K, which guards against excessive bleeding. It is also a good supplement to strengthen the fetus. It can be used during labor to tone the uterus easing the painCan increase milk production in lactating women. Nettles provide nourishing amounts of: Vitamin A, Choline, Vitamin B6, Betaine, Vitamin K, Thiamin, Riboflavin, Folate, Niacin, Calcium, Magnesium and Iron.[xviii]

Notes

Ginger

Parts Used: rhizome (underground stem) fresh, powdered, dried, or juice.

Systems Supported: digestive and immune

Ginger is an herbal <u>stimulant</u>. A stimulant warms the body and increases metabolism an circulation.

Ginger is an anticatarrhal. An anticatarrhal eliminates or decreases the production of mucous.

Ginger is <u>anti-inflammatory</u>. An anti-inflammatory reduces inflammation.

Ginger is a <u>carminative</u>. A carminative relieves intestinal gas and bowel pain. (Gripe)

Ginger is a <u>sialagogue</u>. A sialagogue increases saliva to aid in digestion, specifically of starches.

Ginger is a <u>diaphoretic</u>. A diaphoretic induces sweating.

Ginger is so useful in alternative care and effects so many varied body systems it almost does not seem as if it could be true. However, since many chronic diseases are caused by undiagnosed, underlying gut inflammation, it can be understood how the daily consumption of ginger can effect overall health to such a great degree. Commonly ginger is used to stimulate digestion by increasing the actions of the stomach and small intestines. Aids in indigestion and intestinal cramps by increasing the circulation through stimulation. Ginger reduces inflammation, particularly inflammation in the gut.It is excellent when taken with lemon and honey for colds and the common flu. Ginger clears microcirculatory systems including the sinus and aids in throat and nose congestion. Ginger promotes healthy sweating when taken during the common cold. German researchers have found dermicidin to be a microorganism produced in the sweat glands, that when secreted in sweat is transported to the surface of the skin and aids in germ death. According to studies published in the Lancet, ginger is beneficial in reducing both morning and motion sickness related nausea. A review of six double-blind, randomized controlled trials with a total of 675 participants, published in the April 2005 issue of the journal, Obstetrics and Gynecology, has confirmed that ginger is effective in relieving the severity of nausea and vomiting during pregnancy. The review also confirmed the absence of significant side effects or adverse effects on pregnancy outcomes. Ginger contains two major enzymes: COX (cyclooxygenase) and LOX (5-lipoxgenase) These two enzymes are known to treat chronic inflammation. Most common drugs used to treat diseases, including arthritis, block COX, but not LOX. Plus, these drugs are so powerful, they commonly have ulcerous side effects. By blocking COX altogether they exclude the benefits COX has on our digestive system. COX helps in the absorption of essential nutrients. What is inflammation? Inflammation is the body's natural healing response to illness or injury, and its pain, redness, heat, and swelling are attempts to keep you from moving a damaged area while it is being repaired. Inflammation subsides as the body heals. Ginger has been shown to be more effective against staph infections than some antibiotics. Topically, ginger is used for joint pain or inflammation, including muscle pain among many other uses.

Warnings:

*Ginger prevents clotting and thereby improves cardiovascular function; however caution should be used prior to labor or surgery if medicinal quantities of ginger are used daily due to this factor. Also, do not take when taking blood thinners.

Notes_____

> *"The highest ideal of cure is the speedy, gentle and enduring restoration of health by the most trustworthy and least harmful way."*
> — Dr Samuel Hahnemann

Garlic

Parts Used: bulb

Systems Supported: digestive, immune, circulatory

It is considered to contain <u>antibacterial</u>, <u>antiviral</u> and <u>antifungal</u> properties.

Part of the onion family including shallots and leeks.

Allicin is the vital compound found in garlic, along with ajoene, alliin and sulfur

Garlic is a source for: Vitamin A, Vitamin B complex, Vitamin C, Calcium, Zinc, Selenium

Increase blood motility through arteries thereby decrease heart disease

Prevents food poisoning Aids in treating ringworm, bacterial vaginosis and, yeast. Garlic is one of the best known broad-spectrum antibiotics in the natural world. Garlic can help to prevent food borne bacteria and food poisoning by killing E-coli and Salmonella, two of the most common food borne pathogens. It has also been shown to kill the ulcer causing bacteria heliobacterpelori. When consumed with chlorophyll the odiferous effects of garlic or negated. Try eating garlic with parsley or other green, leafy vegetables. 1mg of garlic has the potency of 15 SU (standard units) of penicillin. Most US diets are deficient in sulfur, of which garlic is an incredible source. Sulfur is the third most abundant mineral in your body, after calcium and phosphorous. Sulfur bonds are required for proteins to maintain their shape, and these bonds determine the biological activity of the proteins, and is found in muscles, skin and bone. Garlic improves iron metabolism, so be sure to consume garlic along with iron for better absorption. Garlic protects the heart by maintaining the elasticity of the heart vessels as we age. It is able to do so by inhibiting angiotensin iii which is a protein that causes vessel constriction. Studies have shown garlic to equalize blood pressure so whether you have low or high BP regular consumption of garlic can help regulate your pressure. Garlic promotes expectoration. Chopping and crushing the garlic releases the vital enzymes from the allicin increasing the effectiveness. For best effect, chop and/or crush garlic and allow to sit before consuming. Garlic has also been shown to kill adult roundworm, one of the most common parasites to effect humans. In a 12-week double blind study garlic was shown to decrease the length of a cold by 50%. Research from gynecologists at the Chelsea and Westminster Hospital in London showed that taking garlic during pregnancy can cut the risk of preeclampsia. Another study showed consuming garlic helped to increase the birth-weight of babies thought to be SGA.

Rosemary

Parts Used: leaves

Systems Supported: immune, nervous, integumentary and circulatory.

Rosemary is a stimulant.

Rosemary is a vulnerary herb. Vulnerary herbs promote cell growth and encourages wound repair.

Rosemary is an astringent. An astringent is a substance which has a binding or constricting effect.

Rosemary is a diaphoretic. (induces sweating)

Rosemary is an antibacterial. An antibacterial inhibits the growth of bacteria and harmful amoebas.

Rosemary is part of the mint family.

Rosemary is used in treating headaches and can replace aspirin or Tylenol, especially when combined with peppermint. A good source of iron, calcium, potassium, manganese, copper, magnesium, Vitamin A, Vitamin C and Folic Acid and other vital B Vitamins including 6. Rosemary can be infused into an oil and used externally for skin irritations like eczema and joint problems like arthritis, and has also been reported to speed healing of wounds and bruises when used externally. It is used topically in treating scalp and hair conditions Rosemary infused oil is an intensive treatment for bad dandruff or fungal related hair loss and can improve scalp conditions! One of the most commonly smoked herbs, rosemary is primary in relieving asthma and mucous related lung conditions. Research has found that rosemary contains a diterpine called carnosic acid that has neuroprotective properties that researchers believe may protect against Alzheimer's disease as well as the normal memory loss that happens with aging. Remarkably, even the smell of rosemary has been found to improve memory. Rosemary has aided in liver detoxification and health since the days of Hippocrates. There are few human studies done on rosemary, but research in mice have shown, rosemary reduces cirrhosis when toxins are present in the liver. Several studies show that rosemary inhibits foodborne pathogens like Listeria monocytogenes, B. cereus, S. aureus., H. pylori (the bacteria that causes stomach ulcers) and Staph infections.

Warnings:

Generally considered safe in recommended does, rosemary has many drug interactions and should be used with caution if taking medication. Rosemary should not be consumed therapeutically by pregnant women. Also, people with high blood pressure should not take rosemary because it tends to raise BP in some instances.

Thyme

Parts Used: Leaves and flowers
Systems Supported: Respiratory, digestive and immune

Thyme is an antiseptic. An antiseptic weakens and slows the growth of bacteria and other microorganisms, which in turn helps to prevent the bacteria from causing further infection
Thyme is an antifungal. Antifungals inhibit the growth of fungus.
Thyme is a diuretic.
Thyme is a vulnerary. (promotes cell growth and encourages wound healing.)
Thyme is a rubefacient. Rubefacients increases blood flow to the surface area of the skin where applied. Their function is to draw inflammation and congestion from deeper areas of the body.

Thyme is an <u>antispasmodic</u>. Antispasmodics prevents or relaxes muscle spasms.

Thyme contains vitamin K, iron, manganese, calcium, selenium and dietary fiber. Thymol—named after the herb itself—is the primary volatile oil constituent of thyme, and its health-supporting effects are well documented. Used extensively to relieve sore throats, coughs and bronchial related concerns. Both thyme and basil have antimicrobial effects on food born illness. Treatment of parasites both internal and external, including lice. Use as a rinse for styes or pink eye. Place 200 grams of thyme in a bath. Someone who is suffering from over stimulation or depression would use this remedy. The thyme bath brings immediate relief and brings on restful sleep. Leeds Metropolitan University in England found that thyme was effective at fighting the bacteria that causes skin acne. Herbs like thyme are thought to be gentler on the skin than other products because of their anti-inflammatory properties.

Mint

Parts Used: leaves
Systems Supported: digestive

-Peppermint is <u>carminative</u>. (relieves gas)
-Peppermint is a <u>stimulant</u>.
-Peppermint is <u>antispasmodic</u>.
-Peppermint is <u>diaphoretic</u>.

Menthol is the most active chemical found in mint. Peppermint brings relief to: stomachache, flatulence, soothes gallbladder pain, soothes IBS (a collection of symptoms including abdominal cramping and pain, bloating, constipation, and diarrhea) prevents bacterial growth, in the small intestine, relieves mornings sickness and other forms of nausea and relieves cough. Peppermint has been used to boost energy. The chemicals found in mint have shown to relax the smooth muscles in the walls of the intestines. Peppermint thins mucous and breaks up phlegm. Used in the reduction of infantile colic. Peppermint infused water can be used to combat breastfeeding ailments such as cracked or damaged tissues. Headaches can be relieved by using peppermint essential oil topically.

Basil

Parts Used: leaves

Systems Supported: Digestive, respiratory and nervous.

-Basil is <u>antipyretic</u>. This means it is a cooling herb and works well to eliminate or reduce fevers.
-Basil is a <u>carminative</u>.
-Basil is a <u>stimulant</u>.
-Basil is a <u>diuretic</u>.
-Basil is a <u>nervine</u>. These are herbs that calm nervous tension and nourish the nervous system.

Much like thyme, basil can be used as fruit and vegetable wash to prevent food borne illness caused by harmful bacteria and amoebas. A tea made of basil and crushed black peppercorns is an effective aid in reducing eruptive and malarial fevers. Basil tea is excellent to reduce phlem and lesson coughs. Boiling basil leaves with honey and ginger is useful for treating asthma, bronchitis, cough, cold, and influenza. Boiling the leaves, cloves, and sea salt in some water will give rapid relief of influenza. Basil infused water can be used as a gargle to reduce sore throat pain and clear bacterial infections. Boiling basil leaves and taking 1 tbs per hour can lessen a headache. Basil is one of the best nourishing herbs as it contains copious amounts of: Vitamin K, Iron, Calcium,Vitamin A, Fiber, Manganese, Tryptophan, B6, Magnesium, Vitamin C, and Potassium.

Oregano

Parts Used: leaves and stems

Systems Supported: respiratory, immune, urinary and integumentary

Oregano is anti-inflammatory, antiseptic, carminative stimulant antispasmodic and antibacterial.

Oregano is incredibly nourishing and balancing when it comes to the daily intake needs of vitamins and minerals and includes: excellent sources of potassium, calcium, manganese, iron, magnesium, carotene, Vitamin C, Vitamin K, folate and dietary fiber. Oregano infusion is taken by mouth for the treatment of colds, influenza, mild fevers, indigestion, stomach upsets, and painful menstruation. The active principles in the herb may improve the gut motility in addition to increase digestion by facilitating gastrointestinal enzyme secretions.

Catnip

Parts Used: leaves

Systems Supported: nervous and digestive systems

Catnip is one of the mildest herbs and will be discussed more thoroughly during the section on children's herbs, however, it must be noted in the nourishing herbs section as well. Catnip can be taken daily, specifically as tea. It is a calming herb that can be used both internally and externally and is overall a nourishing herb for our nervous system.

Elderberry

Parts Used: berries

Systems Supported: immune

This is a nourishing botanical that should be taken daily beginning in the fall months and throughout the winter. It is immune building botanical that helps to strengthen weakened systems. It can be taken daily year round, but most certainly should be part of any healthy regiment to remain strong during the common cold and flu months. Elderberry has been clinically proven to prevent eight types of influenza. The complex sugars of the berries activate immune-activity and disarms harmful enzymes preventing viruses from penetrating healthy cells specifically in the lining of the nose and throat. Elderberries also contain pigments, tannins, amino acids, carotenoids, flavonoids, sugar, rutin, viburnic acid, vitaman A and B and a large amount of vitamin C. They are a mild laxative, a diuretic, and diaphoretic. Research shows elder to stop the production of hormone-like cytokines that direct a class of white blood cells known as neutrophils which cause inflammation and increases the production non-inflammatory infection-fighting cytokines as much as 10 times.

Hibiscus

Parts Used: flower petals

Systems Supported: immune

Hibiscus tea can be added to all nourishing herb blends and formulas and can be taken daily. It is high in vitamin C and is known to reduce free radicals in the body. It stabilizes blood pressure and supports the cardiovascular system. Hibiscus is also know to eliminate bacterial infections and can be particularly helpful to both the respiratory and urinary systems.

Hibiscus sabdariffa also called Rosella
Hibiscus is a diuretic.
Hibiscus is a choleretic which is a substance that increases secretions of bile.
Hibiscus is also considered to be antispasmodic, anthelmintic (expels parasites) and antibacterial.

This plant is part of the mallow family It originated in Egypt and can now be found growing in warm places around the world including India, Africa, Sudan, Jamaica, China, Philippines, and the United States. All the parts of Hibiscus Sabdariffa L. are used it is traditionally known to be a <u>laxative</u>, <u>diuretic</u>, <u>anti-bacterial</u>, and because of its high vitamin C content, <u>antiscorbutic</u> which protects against scurvy. High in Vitamin C, it could be a great choice of beverage to reach for when fighting a cold and flu. Hibiscus tea is also known as sour tea and and is very tart tasting, and can be used as a liver tonic. Hibiscus syrup or honey should be used to reduce cough.[xix]

Lemon balm

Parts Used: leaves

Systems Supported: nervous and digestive systems

Lemon balm is a gentle building herb. Like all nourishing herbs, they are slow and steady in the work they do. Lemon balm can be taken daily, with the contraindication for those who are lactating or who have thyroid disease. Lemon balm, as a part of the mint family, can be used in combination with many other nourishing herbs. They can calm, ease and aide our bodies overall well-being. Lemon balm is considered: anti-bacterial, anti-oxidant, anti-spasmodic, anti-viral, <u>carminative</u>, <u>cerebral stimulant,</u> <u>diaphoretic</u>, <u>digestive</u>, <u>emmenagogue</u>, <u>nervous restorative,</u> <u>mildly sedative</u> and a <u>tonic</u>. Lemon balm is one of the first herbal defenses against upset stomach as it relaxes smooth muscles of the intestines. Lemon balm is also known to help in the following: amenorrhea, anxiety, calming nerves, chronic fatigue, colds, colic, depression, gastrointestinal complaints, Graves' disease, headaches, hypothyroidism, insomnia, menstrual cramps, nausea relief, nervous agitation, palpitations, shingles, and can be used as a wound wash Mixed with feverfew, lemon balm is highly effective for migraine prevention.

Turmeric.

Parts Used: rizommes

Systems Supported: immune and digestive systems

Turmeric is an anti oxidant and a cell builder. Always good to have a cell builder in your daily nourishing routine. Turmeric can be used daily in all manner including: smoothies, teas, cooking, tincture and can also be used externally. Turmeric also is a known anti-inflammatory one of the leading causes of chronic illness today. Taken daily turmeric has endless benefits.

-Turmeric is part of the ginger family
-It is considered today to be the super-spice and is even being embraced my allopathic doctors as well as herbalists.

The medicinal benefits of turmeric include:

-Reduction in inflammation
-Treatment of wounds
-Considered to delay Alzheimer's
-Supports liver and gall bladder function.
-Helps support overall well-being and decreases depression.
-Relieves heartburn and aids stomach ailments
-Chemicals found in turmeric show to significantly decrease systemic inflammation.
Just 2 tsp of turmeric contain our daily manganese requirement and 10% of our iron needs. It also contains: dietary fiber, a complex of B Vitamins along with Vitamins C, E, K and potassium, iron, zinc and sodium. Daily consumption is considered to prevent anemia.

Humans make thousands of units of vitamin D within minutes of whole body exposure to sunlight.

From what we know of nature, it is unlikely such a system evolved by chance.
Dr. John Cannell

Notes_____

_____
_____
_____
_____
_____
_____
_____
_____
_____
_____
_____
_____
_____
_____
_____
_____
_____
_____
_____
_____
_____
_____
_____
_____
_____
_____
_____
_____

The garden is the poor man's apothecary.

Herbs for the Body Systems

Herbs have three functions:
1. _____
2. _____
3. _____

Circulatory System

Hawthorn

Garlic

Motherwort

Butcher's Broom

Ceyenne

Nervous System

Skullcap

Hops

Catnip

Kava

Lemon Balm

Lavender

Passionflower

Valerian

Lymphatic System

Garlic
Echinacea
Golden Seal
Yarrow
Yellow Dock

Respiratory System

Eucalyptus
Lobelia
Peppermint
Angelica
Hyssop
Thyme
Mullein
Rosemary
Oregano
Sage

Immune System

Turmeric
Echinacea
Garlic
Ginger
Elderberry
Aloe
Nettles

Licorice

Hyssop

Astragulas

Cats Claw

Golden Milk

1 1-inch knob fresh turmeric

1 1/2-inch knob fresh ginger

1 cup coconut milk

½ cup water (optional)

1 tablespoon raw honey

Peel turmeric and ginger and grate with a zester. Pour the coconut milk and water into a saucepan add the ginger and turmeric paste. Heat until just boiling. Allow to steep. Strain if desired and add honey to taste. Serve warm.

Turmeric Smoothie

1 cup coconut milk
1 teaspoon grated turmeric
1 ripe banana
Know of grated ginger
Chia seeds
½ teaspoon cinnamon
Honey, to taste 1
Vanilla to taste

Turmeric Juice

1 large knob of turmeric

1 large knob of ginger

1 apple

1 orange

1 lime or lemon

1 carrot

1 pinch of cinnamon

Place all of the above listed ingredients in a juicer or a blender. Mix and drink daily.

Echinacea

Echinacea is a member of the sunflower family. It is known to support infectious disease suppression and poor immune function. Also used to treat influenza, colds, chronic fatigue syndrome, and aids through immune system support. Echinacea stimulates natural immune function by increasing white blood cell activity and raising the level of interferon's, to resist invading microbes. Echinacea is a compound in Echinacea, which prevents germs from penetrating healthy cells, suppressing viruses and bacteria in the body. Echinacea can be used topically or internally to treat and prevent infections. There are many ways to use elderberry, but one of the best ways is to make a tincture.

How to Make a Tincture

-Tinctures are highly concentrated herbal extracts that can be kept for long periods of time.

-Tinctures are made with high quality alcohol or vinegar.

-Combine herbs in a sterilized jar and cover with alcohol or vinegar.

-Keep in a cool dark place and shake daily.

-Allow the herbs to extract 2-6 weeks.

- The amount taken varies from herb to herb but often is from 10-20 drops

- Tinctures are often added to water or juice to make more palatable

- This amount of alcohol is miniscule, but if alcohol is not to be consumed at all, make with vinegar instead.

- Tinctures made with vinegar lasts 6mths.

- Tinctures made with alcohol will last indefinitely

- Tinctures are used primarily for poor tasting herbs or herbs needing to be taken over long periods of time.

Marseilles Vinegar

2 Tablespoons Thyme

2 Tablespoons Rosemary

2 Tablespoons Cloves

1 Tablespoon Black Peppercorns

2 Tablespoons Sage

2 Tablespoons Lavender

2 Tablespoons Mint

¼ cup Elderberry

8 Cloves Chopped Garlic

¼ Chopped Onion

1 Knob Fresh Ginger Grated

1 Knob Fresh Horseradish Grated

One 32-ounce bottle of organic Apple Cider Vinegar with the Mother

Used dried herbs. Combine all ingredients and place in a dark cool place. Let stand 6-8 weeks. Shake daily. Strain. Take a minimum of 3 tablespoons per day.

Urinary System

- Elder flower
- Juniper
- Lady's mantle
- Plantain Leaf
- Ginger
- Grape seed Extract
- Yarrow
- Uva Ursi
- Rosemary
- Garlic
- Nettles

Integumentary System

The largest organ we have is the skin. This system also includes your hair and nails.

- Chamomile
- Aloe Vera
- Burdock
- Lemongrass
- Horse Tail
- Calendula
- Lavender
- Peppermint

Gotu Kola
Nettles

Muscular and Skeletal

When using herbs for the muscular and skeletal systems we must understand the cause of inflammation and pain leading to disability. To understand this you must consider the fluids and tissues, as well as the structure. Matthew Wood lists the following as herbs for the muscular and skeletal systems:

Cerebrospinal Fluid: black cohosh
Synovial Fluid: pleurisy root, comfrey.
Interstitial Fluid: Gravel root, boneset, mullein.
Blood Vessels and Nerves: St. John's Wort, prickly ash.
Muscles and Bones: boneset, comfrey, mullein, rue.

Digestive System

As you begin explore herbs, more and more you will find the most common action amongst many herbs is related to aiding digestion. Early on, I wondered at this. Why was the main action of most herbs to help digest food, soothe irritated stomachs and reduce inflammation? The more I have learned and come to understand how the body works, and how chronic illness is able to pervade and invade our lives, I understand why most all herbs aide digestion. It is the action we need most. As modern food progresses, or rather digresses, it is paramount we support our systems with herbs which soothe our intestines and increases gastric digestive processes. Altogether our intestines are about 25 feet in length. This means inside our bodies we have 25 feet of an unseen area that is susceptible to consistent inflammation. In the end this could well be one of the leading reasons for chronic diseases we face.

Herbs and botanicals for digestion include: aloe, ginger, cinnamon, fennel, lemon balm, lemon grass, peppermint, burdock root, dandelion, papaya, chamomile, orange clove and cardamom. Mint can help some digestive conditions, like indigestion and gas, yet cause harm in others such as heartburn or acid reflux.

The oil in the peppermint relaxes nerves and improves circulation.

Tincture can be taken for nausea in pregnancy.

Menthol and methyl salicylate are the main active ingredients of peppermint, and have anti-spasmodic actions which calms muscles spasms of the stomach and calms the intestinal tract, and uterus.

These oils also have analgesic properties reducing pain associated with the stomach.

Dandelion

Parts Used: Leaves, flowers and roots

Dandelion is a diuretic, antacid, stimulant and a tonic specifically for the liver.

-Of all lowly weeds dandelion is one of the lowest, however, it is also touted as one of the greatest herbs by those who know.

-It is one of the top 6 herbs in the Chinese herbal medicine chest.

-Dandelion has been used to calm upset stomach, intestinal gas and dissolve gallstones.

-This weed-like superfood helps the kidneys clear out waste, salt, and excess water, and it also inhibits microbial growth in the urinary system too.

-Dandelion is also used to increase urine production and as a laxative to increase bowel movements.

-It is also used as skin toner, blood tonic, and digestive tonic.

Cinnamon

Parts Used: Bark (sticks/quills)

Cinnamon is an anti-fungal, and antibacterial.

Cinnamon is a stimulant, demulcent and caminative.

Ceylon is known as true cinnamon and originates in Sri Lanka

Cassia is known as common cinnamon or Chinese cinnamon and is most likely found in the stores in the West.

Cinnamaldahyde is the main chemical in cassia cinnamon.

Cinnamon is known to increase digestion.

Cinnamon also is said to prevent blood clots.

It is used to combat intestinal bloating and stomach upsets, including diarrhea and vomiting.

Cinnamon has also been known to help alleviate stomach upset related to the stomach flu.

Cinnamon is a warming herb and should be used with caution if pregnant.

Cinnamon Tea is to be taken at the first sign of a cold or flu.

Cinnamon Tea

Boil 3 cups water
Steep the following:
10 cloves
4 cinnamon sticks
2-3 cardamom pods (optional)
Knob of ginger
2 bags of tea
Cayenne (optional)

Add for taste:
Lemon
Honey

Cinnamon aids in digestion. Drink this tea following a meal. This tea also soothes sore throats and can help alleviate coughs. You can add milk to taste, but not if drinking for a cough. Fennel seeds are used to treat digestive ailments. The seeds are a carminative, which means they help prevent or remove gas from the intestines. Fennel seeds are a common ingredient in "gripe water," which is gentle enough and can be used to treat colic in infants.

Fennel seeds help alleviate: heartburn, intestinal gas, bloating, loss of appetite, upper respiratory tract infections, coughs, bronchitis, cholera, backache, bedwetting, and visual problems.

Fennel is used as medicinal treatment for several types of health issues including menstruation.

Fennel is also used to increase the flow of breast milk during lactation.

Fennel Seed Tea

Crush fennel seeds and steep in boiled water with peppermint.

Aids stomach upset.

Add honey to taste.

Lemongrass

Parts Used: Stems and Leaves

Systems Supported: Digestive, Respiratory

Lemongrass is an antispasmodic, antiseptic, anti-fungal, antibacterial, antipyretic.

The main chemical component found in lemongrass is citral, also known as lemonal.

Lemongrass is key to digestive health and aids in preventing heartburn.

Lemongrass also aids in soothing stomach upset and quiets cramping intestines.

Widely used to alleviate certain respiratory conditions including laryngitis and sore throats lemongrass is inhaled through vaporized steam.

Also known as fever grass in many Asian cultures, it has earned this reputation due to the <u>anti-pyretic</u> property which reduces high fevers. Again, inhaled through steam perspiration is increased and the fever is soon break. Lemongrass also has powerful pain relieving properties. It helps to alleviate muscle spasms by relaxing the muscles thereby leading to the reduction of pain-related symptoms. It is thus useful for all types of pain including abdominal pain, headaches, joint pains, muscle pains, digestive tract spasms, muscle cramps, stomachache and Found in the carrot family. This family also includes celery, parsley, cumin and parsnip.

Cilantro and Coriander Seed

Parts Used: Coriander is the seed of the cilantro plant; Cilantro is the green leafy tops.
Systems Supported: Digestive

An infusion made of cilantro can be used to wash eyes and relieve conjunctivitis.

In Europe coriander is known as the diabetic plant and is utilized to lower blood sugar.

Coriander consists of high levels of dodecenal which has been shown to be twice as effective as in treating salmonella borne illnesses than the common prescribed antibiotic.

Cilantro wards off flatulence and aids in digestion. Cilantro is also the best known herb for chelation and is known to remove heavy metals including, mercury, aluminum and lead.

Papaya

Parts Used: Fruit and Leaves

System Supported: Digestion

-Enzymes in papaya leaves can break down indigestible foods in the gut and intestines.

-The riper the fruit, the less enzymes.

-Papaya breaks down proteins carbs and fats

-Carpain, the chemical found in green papaya, is able to kill parasites.

-Papaya is not generally used topically as it can cause irritation.

It contains similar substances to latex which is derived from the fig tree.

-Medicinal amounts taken while pregnant could cause an abortion or birth defects.

-3 leaves per day, made into a tea, is considered a therapeutic dose.

-Chymopapain, also a chemical found in papaya, aids in digestion.

-Papaya Leaf contains; Vitamins A, C, E, K, B Complex and is especially high in B17

-Papaya leaf has been said to aid: bloating, chronic indigestion, intestinal worms, increases immune function, decreases inflammation, promotes the healthy production of cytokines which also works to regulate the immune system,

-Papaya seeds can reportedly kill: E-coli, salmonella, Staph, and other bacterial infections.

Cardamom

Cardamon is an anti-fungal, antibacterial, antispasmodic and antiviral. It is also known as an overall tonic.

Cardamon aids in digestion and prevents heartburn, intestinal spasms, irritable bowel syndrome (IBS), intestinal gas, constipation, liver and gallbladder complaints, and loss of appetite.

Cardamon pods can be chewed to alleviate hiccups.

Cardamon helps maintain oral health.

Taken daily as a tonic it is thought to improve

overall well-being and reduce depression .

Add cardamon to cinnamon tea for a potent digestive tea. Flavor with honey.

Clove

Parts Used: unopened flower buds of the evergreen clove tree.

The essential oil of cloves contain eugenol and beta-caryohpyllene which produce a mild anesthetic often used in oral health and gum, mouth and tooth pain.

Clove is often used in throat spray and mouth wash recipes because of the antibacterial/anti-fungal effect on germs.

Clove is also considered to be anti-inflammatory, antiseptic and anti-parasitical.

Cinnamon and clove tea is considered an anti-fungal with mild antiseptic and used in the case of a sore throat.

Cloves also aids in digestion.

Cinnamon and clove tea is taken to reduce sore throat pain.

Chai Tea

1 whole fresh ginger rhizome grated

10 cinnamon sticks

2 tablespoons black peppercorns

10-20 whole cloves

10 cardamom pods (opened and crushed)

8 cups cold water

8 bags of black tea

2 cups whole milk

1 vanilla bean split and spread

1/2 cup raw sugar or raw honey

Add to slow cooker and steep for 3-4 hours and add milk, sugar and honey to taste.

Herbs for Heartburn

Ginger
Caraway Seed
Chamomile
Lemon balm
Licorice
Peppermint
Turmeric
Marshmallow root
(Increased calcium)

Herbs for Acid Reflux

Ginger

Aloe Juice

Raw apple cider vinegar
(Increased folic acid)

"Of all the herbs I've ever studied -- aloe vera is the most impressive herb of them all."
Mike Adams

Notes

Herbal First Aid

These herbs are excellent to add to your herbal first aide kit: **aloe, comfrey, cayenne, calendula, plantain, lavender, and yarrow**

Aloe

- Aloe is an <u>alterative</u>. This means "blood purifier".
- Aloe is a <u>chologogue.</u> Promotes the flow of bile into the small intestines.
- Aloe is a <u>demulcent.</u> Soothing substance, usually taken internally, protects damaged tissue.
- Aloe is a <u>vulnerary.</u> Promotes healing of wounds.

Parts used: Leaves/gel

Systems Supported: Integumentary, Digestive

- Aloe is widely used for sunburns and first degree burns. It both soothes and speeds healing.
- It also can be applied topically to treat insect bites and soothe itchy dry skin.
- Aloe helps maintain elasticity to the skin.
- Taken internally aloe is a laxative. Combining aloe with ginger alleviates bowel cramping.
- Aloe Vera contains many vitamins and minerals including A, C, E, folic acid, choline, B1,B2, B3 (niacin), B6. Aloe Vera is also one of the few plants that contains vitamin B12. It also contains calcium, magnesium, zinc, chromium, selenium, sodium, iron, potassium, copper, manganese.
- Aloe Vera is a well-known <u>adaptogen</u>. An adaptogen is something that boosts the body's natural ability to adapt to external changes and resist illness.
- Disease cannot manifest in an alkaline environment. Most people are living and subsisting on mostly acidic foods. For great health, remember the 80/20 rule – 80% alkaline forming foods and 20% acidic. Aloe vera is an alkaline forming food. It alkalizes the body, helping to balance overly acidic dietary habits

Aloe Salve

You will need 3-5 <u>large</u> aloe vera leaves.

Split the leaves down the middle lengthwise and scrape out the gel and place in into a 1 cup measure.

Fill the rest of the cup with olive oil.

Grate 1 quarter cup of beeswax.

Place the olive oil and aloe gel into an old pan over low- medium heat incorporate bees wax until melted.

Test for consistency by placing a small amount of salve on a spoon and set in the refrigerator for a few minutes to cool. If the salve is too runny, add a little more beeswax. If too stiff, add a little olive oil.

Comfrey

Parts used: Leaves and roots

Externally, comfrey is said to be <u>anti-inflammatory</u> and speed wound healing due to its natural concentration of allantoin.

Comfrey is a <u>vulnerary</u> and is an <u>astringent</u>.

Comfrey can be used for minor injuries of the skin, where it will work to increase cell production, causing wounds to heal over rapidly. It will help reduce inflammation, and at the same time lessen scarring.

Comfrey is also known as knitbone and can be used when a bone is broken to speed healing and repair.

Herbal Compress

Bug Bites: Basil, Plantain
Mild Burns: Peppermint, Sage, Eucalyptus, Chamomile, Green Tea, Rose
General Skin Irritation: Plantain, Chamomile, Calendula, Lavender, Rose, yarrow
Sprain: Comfrey, calendula

Yarrow

Yarrow is one of the oldest documented herbs dating back to before Egyptian society. It's claim to fame numerous, including being a woman's herb. It's wound healing properties are matched by few. It's next greatest claim to fame is it's ability to make us sweat. When fever is building, drinking hot teas of yarrow can help it to break by relaxing the circulation and open the pores of the skin, allowing us to sweat freely and ridding the body of infection.
Use as: teas, poultices, spit poultices, steams, tinctures, oils, and vinegars.

Plantain

Plantain is one of the most common weeds in the world, and is a go-to remedy for many skin infections, rashes and wounds. It is safe and effective for: bee stings, bleeding, cuts, bruises, bug bites, hemorrhoids, and itchy skin. Make a "spit" poultice by chewing the leaf or in a mortar and pestle. The leaves can also be made into a tea or tincture, and this is said to help with indigestion, heartburn and ulcers when taking internally.

Calendula Flower

Calendula flower is used to prevent muscle spasms, start menstrual periods, and reduce fever. It is also used for treating sore throat and stomach and duodenal ulcers.

Calendula is applied to the skin to reduce pain and swelling or inflammation and to treat poorly healing wounds and leg ulcers. It is also applied to the skin for nosebleeds, varicose veins, hemorrhoids, and inflammation of the lining of the eyelid. Flowers are applied to cuts and wounds to stop bleeding, prevent infection and speed healing.

Feverfew

Feverfew inhibits clotting and inhibits the blockage of small capillaries. Feverfew has a mild tranquilizing effect which relieves headaches caused by tension. Feverfew is also used to reduce inflammation, menstrual discomforts, and fever.
Use as: Skin wash, tincture and tea.

Hops

Use for:

Pain

Anxiety

Sleeplessness or Insomnia

Bruises

Sprains

Hormone regulation

Studies have found that hops promotes sleep through sedation. The chemical compounds in hops both relieves pain and kills bacteria. The phytochemicals in hops are considered tonic in nature and possess nervine and muscle-relaxing qualities which depress the central nervous system.

Poultice

Comfrey Leaf and Root
Lobelia

Marshmallow Root

Mullein Leaf

Turmeric

All aide in relieving sore muscles.

Cayenne

-Cayenne is a <u>stimulant</u>, <u>astringent</u>, <u>carminative</u>, and <u>antispasmodic.</u>

-It is considered a superior herb for crisis situations and in first aid related incidences.
Parts Used: fruit and seeds
Systems Supports: circulatory, immune, digestive
-Cayenne can be taken as a daily tonic to help maintain a healthy cardiovascular system and prevent strokes, heart attack, colds headaches, and joint inflammation
-Cayenne powder or tincture can be rubbed on
toothaches.
-Cayenne oil, tincture or cream can be rubbed on aching muscles or joints to aid in pain relief.
-Since cayenne normalizes circulation it has been relied on to arrest internal or external a hemorrhage
-Combined with ginger, lemon and honey cayenne has been used to aid sweating and stimulate weakened internal organs.
Cayenne is a <u>thermogenic</u>. It raises the core temperature and stimulates metabolism and the production of endorphins, or the feel-good transmitters, and creates a sense of well-being. Routine consumption of cayenne reduces oxidative stress and free radical damage.
-Cultures where cayenne is highly valued in the diet report lower rates of heart attack, stroke and pulmonary embolism
-Internally capasicin, the main chemical compound found in cayenne, reduces triglyceride levels and platelet aggregation and increases the body's ability to dissolve fibrin which causes the formation of blood clots.
-Cayenne contains essential nutrients vital to the development of healthy mucous membranes. These membranes line the nasal passages, lungs, intestines and urinary tract.
-Cayenne has also been used as a method of alleviating headaches.

Warning: Excessive consumption of cayenne has shown to cause stomach and esophageal erosion. Caution must always be taken when ingesting herbs.

Consumption and Uses:

-Cayenne powder for first aid kit
-Cayenne tincture
-As an additive to tea.
-As a topical oil or cream for pain relief

Infused Oils

Choose the herb. Here we are choosing cayenne which will be used as a massage oil for stiff joints or sore muscles.

Wash and DRY THOROUGHLY (cayenne does not have to be washed). Never use an herb with moisture left inside especially when oil will be consumed. The deadly botulism can, and will grow, in oils where water is present.

Place in a crock pot on low or on the stove on low heat

Bacteria does not generally form in plain olive oil, but it can when water is present as in garlic or wet herbs

Do not boil!

Keep oil on very low heat and allow the herbs to infuse into the oil.

Sterilize and DRY jars before placing herbal oils inside.

Use within 1 week if using internally. Can last for several months when using topically.

Salve

A salve is an herbal preparation held in place, when applied to the skin, due to its thick consistency. One of my favorite salves is cayenne salve for muscle or joint pain relief. This is a warming salve and will bring blood to the area.

To make an herbal salve prepare an infused oil then
add bees wax to it until desired consistency is reached.
Use:
8 oz herbal infused oil
1 oz Beeswax
10-20 drops essential oil of choice
Best Herbal Salves

Cayenne Pepper-soothing for joint or muscle pain

Other suggestions to be used as salves:

Chamomile Flowers for: hemorrhoids, minor abrasions, cuts, scrapes, and wounds.

Comfrey Leaf: Relieves pain, swelling, promotes the growth of muscle, cartilage, and bone. Assists with healing a wide variety of conditions including sprains, eczema, dermatitis, viral skin infections, broken bones, arthritis, wounds, and bruises.

Ginger: Warming, use for arthritis and sore muscles[xx]

Herbs for Children

Herbs for children: catnip, chamomile, peppermint, lemon balm, mullein and elderberry

Catnip

Catnip is an analgesic. sedative. diuretic.

Catnip is used for trouble sleeping (insomnia); anxiety; migraine and other headaches; cold and other upper respiratory infections; flu; and gastrointestinal (GI) upset, including indigestion, colic, cramping, and gas (flatulence). It is also used as a tonic, for increasing urination.

Catnip, along with chamomile is one of the best go-to remedies for children's sleep disturbances as it is extremely gentle, not addictive and has no known side effects in the morning. Catnip teas have long been used in traditional herbal medicine to quell digestive disturbances, and reducing the pain of menstrual cramps. A hot cup of catnip tea is excellent for treating colds and flu because it producing perspiration without increasing the heat of the system. Catnip tea is good for curing headaches as well. Catnip can be used when a child has a high fever, as it is known to relax the body while increasing perspiration, which helps the infection leave the body faster. It has also been found to settle the stomach and sooth children when they are upset. Externally, I've used Catnip infusions and baths to help with the achy muscles that accompany flu and illness. It can be used externally on the stomach of colicky babies to relax the stomach and help them sleep.

Teething

Chamomile tea is an easy remedy for teething pain. Soak clean cloth in chamomile tea and freeze. Allow baby to chew on frozen cloths. The coolness will sooth sore gums and chamomile can help to soothe mood. Catnip, peppermint, and rose hip teas are also gentle enough to use on a small child.

Chamomile

Chamomile is anti-inflammatory, antimicrobial, antioxidant and antibacterial in nature. The antibacterial effects of drinking chamomile tea can help to prevent and treat colds while protecting against bacterial-related illness and infection.

This herb can:

Calm Muscle Spasms – One study from England found that drinking chamomile tea raised urine levels of glycine, a compound that calms muscle spasms.

-Soothes Stomach Ache – Further adding onto chamomile benefits, the herb is a wonderful for soothing an upset stomach. Helping to soothe and relax the muscles and lining of the intestines, chamomile can help with poor digestion and even those suffering from irritable bowel syndrome (IBS).

Promotes Sleep – Drinking chamomile tea soothes the nervous system so that you can sleep better. It has been used as a solution for insomnia for centuries.

Promotes Healthy Skin – With it's anti-inflammatory and anti-septic properties, chamomile helps in clearing up skin irritations such as eczema, acne, and allergies.

Treats Cuts and Wounds – because of its antibacterial properties it can be used as a wound rinse.

Chamomile is known as a 'tisane'. A tisane is any non-caffeinated herbal concoction made by pouring hot water over the leaves, stems, and roots of plants. You can make your own chamomile tea with other plants like lavender or mint to vary the flavor, or drink it alone.

Earache

Garlic Infused Oil
Onion Infused Oil (or fresh)
Cinnamon Infused Oil
+
Lymphatic
Drainage
Massage

Restlessness

So many children are labeled as ADHD, restless or hyperactive these days. Many children honestly do have problems sitting still and for many reasons. One of the main reasons is that children were not designed to sit still. They were made to run and jump and climb and swing. When we force them to sit in a school classroom then we are setting them up for failure. The second reason a child would tend towards hyperactivity is the vitamin deficiency they have due to a processed foods diet. Adding herbs instead of medications should be considered as a starting place.

The following will provide nutritive values well as a calming effect:

Tinctures to consider:

California Poppy
Valerian
St. Johns Wort

Teas to use:
Chamomile
Lemon Balm
Catnip

Herbs to use for Chest Rub

Children are known to commonly have chest congestion, especially during the winter months. Herbs can break up mucous and open respiratory passages. To relieve chest congestion you can add essential or infused oils to carrier oils or make a salve and rub on their chest at night.

The following herbs can be used together, or alone, as a chest rub and can also be used as a respiratory steam.

Peppermint
Hyssop
Thyme
Lavender
Rosemary
Eucalyptus

Chest Rub

2 parts coconut or olive oil

1 part beeswax

10-20 drops essential oil per ounce

Warm the oil in a pan on the stove. Slowly stir the beeswax into the oil until it is melted. Do not boil. Add more oil to thin and more bees wax to thicken until the consistency is as you like it. Keep in mind as the salve cools it will harden. If you find your salve too hard, then rewarm and add in more oil. Add in essential oil or infused herb oil prior to cooling.

Colds and Flues

Colds and flues are so very common in the lives of children. Before reaching for the bottle on the store shelf, consider giving herbs a try first. Herbs are not filled with chemicals or coloring as is so much of the store bought cold remedies and will also add nourishment.

Snake Juice Cough Remedy

This is a remedy that will last for a very long time and should be made at the beginning of fall in order to be ready for the cold and flue season of winter.

You will need:

4-8 Lemons
4-8 Onions
Honey to cover
1 cup Brandy
Gallon Jar

Chop the onions rather finely.
Chop the lemons rather finely.
Add the onions and lemons to a sterilized glass gallon jar. Add the brandy and mix. Add the honey until all the onions and lemons are covered. Store in a cool, dark place for 6 weeks. Shake daily. This will last indefinitely if not contaminated.

© GoMidwife 2015

Herbal Elixirs

1 clean (sterilized) pint jar with a lid

Fill herb to top of jar

3/4 pint brandy

1/4 pint honey

Chop herbs and fill the jar, leaving an inch of space at the top. Pour in the brandy until it's half-filled, and then fill the rest up with honey (you might have to pour it and wait, repeatedly, until the honey sinks down enough). Screw on the lid, Label, and place in a cool dark area for 6 weeks.

Shake daily. Strain after 6 weeks and enjoy.

Diarrhea

Diarrhea is common in childhood and can be caused by an imbalance in the digestive system whether it is brought on by an enzyme or vitamin A deficiency or intestinal parasites. Symptoms include uncomfortably frequent, fluid, and excessive bowel movements. Diarrhea is a way for the body to rapidly remove toxins, but then cannot compensate for the rapid removal and becomes quickly and severely dehydrated. Loss of vital nutrients leading to quick and significant compromise.

Herbs to use: chamomile, cinnamon, catnip, and fennel
Make a tea out of individual or combined herbs and have child sip consistently.

There are many astringent herbs that help stop diarrhea. Astringents contain tannins that tighten and contract human tissue. The result is fluid retention that can quickly halt diarrhea and reduce any bleeding. Agrimony, bayberry, comfrey, yellowroot, peppermint, slippery elm, white oak, white willow, black walnut, green tea, red raspberry leaf, and mullein are all astringent and excellent for treating diarrhea, dysentery, and even cholera.

The key when treating diarrhea, dysentery, and cholera is to drink plenty of uncontaminated water, to avoid consuming any additional harmful microorganisms, to avoid sugar and fruit juices that feed parasites, to reduce absorption of toxins into the body, to transport toxins out of the body, and finally to heal the bowels with mucilaginous herbs. Mucilaginous herbs soothe and heal damaged parts of the digestive tract. They also carry toxins out of the body without feeding harmful bacteria. Mucilaginous herbs include: psyllium, slippery elm, and marshmallow. They should be consumed with large amounts of water. Okra pods, also considered a mucilaginous herb, should be thoroughly cooked and chewed well when treating diarrhea. It is also a good idea to slowly replace lost nutrients like potassium and sodium by drinking plenty of alfalfa, nettle, or red clover tea. Other foods that are good when treating diarrhea include boiled rice and oatmeal. Mucilage can be sweetened with honey and dosed by the teaspoonful.

Oral Rehydration Solution (ORS)

Oral rehydration salts can be used for children who are dehydrated due to diarrhea or a flu and can also be used for laboring or postpartum mothers.

Here is the recipe:

6 level teaspoons of sugar and 1/2 level teaspoon of salt dissolved in 1 liter of clean water. Be very careful to mix the correct amounts. Too much sugar can make diarrhea worse and too much salt can be extremely harmful to a child. Making the mixture a little too diluted (with more than 1 liter of clean water) is not harmful.

Other fluids that can be used are:

Breastmilk
Gruels (thin porridge)
Carrot Soup
Rice water

If possible, add mashed banana to improve the taste and provide potassium.

Cough Remedy #1

Hyssop
Thyme
Cayenne
Anise Seed
Horehound
Ginger
Mullein
+
Lemon & Honey

Oxymel

An oxymel is the combination of both vinegar and honey for health benefits. Vinegar has long been known to ease sore throats and coughs as well as balance the body In general, herbal preparations that are vinegar based should be aimed at respiratory ills. To make an oxymel, decoct the desired herb, strain and add ½ cup vinegar and ½ cup honey.

Cough Remedy #2

2 lemons peeled and sliced

1 large knob of ginger, peeled and sliced

1 cup of raw honey

Herbal Honey

Rosemary, sage, thyme, mint, lemon balm, lavender and chamomile all make lovely infused honeys.

You can also use spices like vanilla beans, cinnamon sticks, and star anise.

1-2 tablespoons of dried herbs per 1 cup (8 ounces) of honey.

Instructions

1. Prepare herbs: Herbs should be dry (see safety note, below) and may be in the form of whole sprigs or separated leaves, buds, and petals. Chopped herbs may infuse more quickly, but they may also be harder to strain out. (To dry fresh herbs, use an air or oven drying method, dehydrator, or microwave.)

2. Combine herbs and honey: Place herbs in the bottom of a jar and fill the jar almost to the top with honey. Using a chopstick or other implement, stir to coat the herbs with honey. Top off with more honey to fill the jar. Wipe the jar rim with a clean cloth and cover tightly.

3. Infuse: Let the herbs infuse for at least 5 days. If the herbs float to the top, turn the jar over a few times to keep them well coated. For a more intense flavor, infuse for another week or longer.

4. Strain: Strain the honey into a clean jar. Depending on the volume of honey and herbs and the size of the strainer, you may need to do this in stages.

5. Store: Store the honey in a tightly covered jar in a cool, dry place. It will last indefinitely.

Herbal Syrup

Syrups can be made with honey or sugar, herbs and water.

Syrups begin with a very concentrated decoction. Combine an herb or herb blend with water in a pot, using 2 ounces of herb per quart of water. Set the pot over low heat, bring to a simmer, cover partially, and simmer the liquid down to about half the original volume.

For each pint of liquid, add 1 cup of honey or sugar.

Use 2 cups of sweetener to 2 cups liquid a 1:1 ratio sweetener to liquid or if you prefer it less sweet use a ½ :1 ratio. Extra sweetener can help preservation.

Warm the mixture over low heat, stirring well.

Notes_____
_____
_____
_____
_____
_____
_____
_____
_____
_____
_____
_____

> *"The greatest medicine of all is to teach people how not to need it."*
>
> Unknown

Herbs for Sleep and Anxiety

Valarian

There is no specific dose for Valerian, but it seems to be most effective after regularly taken for two weeks or more.

Gamma-aminobutyric acid (GABA) is one of the main ingredients in Valerianarian and helps sooth neurotransmitter receptors.

Valerian has sedative properties generally used to treat mild insomnia, and has been known to treat anxiety and minor pain.

Because the compounds in Valerian produce central nervous system depression, they should not be used with other depressants.

Valerian is often combined with hops, lemon balm and passionflower.

Sleep Aide Tincture

Add 1/2 teaspoon or 10 drops of each tincture: valerian, hops, passion flower, and chamomile to hot water fresh lemon balm. Add honey and lemon to flavor.

Lemon balm is a member of the mint family and is considered to be a calming herb. Lemon balm is especially useful when it is combined with other herbs such as:

Hops, valerian, passion flower and chamomile and skull cap. Lemon balm is known to calm, reduce anxiety and promote sleep.

Skullcap

Fear and anxiety is pervasive in this day and age. Skullcap is an herb which can help alleviate anxiety. Fear is a focused emotion whereas anxiety is more generalized. A small dose of about 10 drops of tincture Can help with a tension headache brought on through anxiety. A larger dose such as 20-30 drops of tincture can promote sleep. Skull cap is thought to help reduce pain associated with nerve pain such as sciatica. Combined with hops, lemon balm, passion flower and chamomile skull cap is quite powerful.

Passion Flower

Passion is a <u>sedative.</u> An herb that strongly quiets the nervous system. Passionflower is used for sleep problems (insomnia), gastrointestinal (GI) upset related to anxiety or nervousness and generalized anxiety disorder (GAD). Passionflower is also used for hysteria, asthma, symptoms of menopause, attention deficit-hyperactivity disorder (ADHD), nervousness and excitability, palpitations, irregular heartbeat, high blood pressure, fibromyalgia, and pain relief.

Warnings: Passionflower is generally considered to be nontoxic when used in moderation. Do not take passionflower if you are already taking prescription medication for anxiety or depression, as excessive sleepiness has been reported. Depression of the nervous system may result in fatigue and mental fogginess if you take too much passionflower for too long. Start with a low dose several times a day and increase as you learn how you respond to passionflower.

Kava Kava

Kava Kava has been used throughout the South Pacific for generation. It is considered an analgesic, or painkiller, and is also traditionally used to lessen anxiety and promote sleep. Kava Kava helps reduce anxiety relating to social stress and calms anxious thoughts and general nervousness. Kava Kava soothes the same nerve receptors as Valium with fewer side effects and less addiction. Kava Kava should not be taken if liver infection or ailment is involved and should not be taken by alcoholics who are already compromising their liver. Kava Kava can be drank as a tea before Bed to reduce restlessness and offset insomnia

Smelling Salts

Place 10-20 drops of lavender essential oil in 1 cup Epsom Salts and smell when anxious or nervous. Herbs that are mild in flavor and taken internally are commonly taken as teas. Fresh herbs are first bruised prior to use. This action breaks up the tissue structure and releases the volatile oils and phytochemicals.

Sleepy Tea

INGREDIENTS

1 tablespoon dried lemon balm
2 teaspoons dried peppermint
1 teaspoon fennel seeds
1 teaspoon dried rose petals
1 teaspoon dried lavender flowers
2 slices dried licorice root
honey to taste
heavy cream or milk to taste

Chaste berry

Chaste berry is also commonly know as Vitex and has often been referred to as the "woman's herb". Chaste berry regulates the female hormones, and should not be taken longer than 6 months at a time.

Chaste berry has been used for:

Menstrual cycle irregularities

Premenstrual syndrome (PMS)

Premenstrual dysphoric disorder (PMDD)

Menopause symptoms

Fibrocystic breasts

Female infertility[xxi]

Preventing miscarriage in women with low levels of progesterone

Preventing PPH

Expelling placenta

Increasing breast milk

Angelica

Angelica is also known as Dong Quai and is one of the most important Chinese herb for women. Angelica can be used as a daily tonic for women, but should never be taken by a pregnant woman or a woman trying to conceive. Angelica is known to help amenorrhea as well reduce the symptoms of PMS. This herb can also be used for women who have a lower sexual desire, especially as they age and move towards menopause. Angelica regulates the effects of hormones in the body and has been known to help reduce hot flashes.
A part from women specific issues, angelica has been used as a liver tonic and to treat pain

associated with nerves such as shingles and sciatica.

Burgamot

Burgamot is <u>antibacterial,</u> <u>anti-spasmodic</u>, <u>diaphoretic</u>, <u>carminative</u>, <u>antiseptic</u>, <u>emmenagogue</u>. <u>anti-inflammaorty</u> and a <u>nervine.</u>

Parts Used: Leaves and flowers

According to some herbalists Burgamot also known as Monarda is one of the primary healing plants native to North America and has long been used as a primary medicine by Native Americans. A strong tea or small dose of tincture can be used to soothe almost any gastric upset including heartburn, nausea, diarrhea, constipation, stomachache or gas. The leaves and flowers can be used in salves and oils for wounds, sprains, bruises, burns, rashes and other external pains to bring down swelling, eliminate infection, reduce irritation, dull pain and stem excess bleeding. Combine Burgamot infused oil with Mugwort and apply to the lower abdomen to ease menstrual cramps. Burgamot is a key herb in treating recurring or chronic yeast infections. It is also a key component to aiding in the healing of leaky gut syndrome. Bergamot tea is a wonderful tonic for the nervous system and is soothing, calming and very much a natural anti-depressant similar to its cousin lemon balm.

Black Cohash

For generations black cohash was used by the women in America. It is known as a female herb and has been instrumental in alleviating many ails specific to women including: PMS symptoms, amenorrhea, dysmenorrhea, and menopausal symptoms. The use of black cohosh dates back to early America and besides being specific to women as an over all tonic it also helps aches and pains associated with muscles. Black cohash is <u>anti-inflammatory</u> and a mild <u>sedative</u>. The active ingredient in black cohash is called phytoestrogen. This is a plant estrogen and a chemical compound very similar to the hormone produced in the female body.

Drinking black cohash tea as a daily tonic can help to alleviate or eliminate perimenopausal and menopausal symptoms. Although these phytoestrogens bind themselves to hormone receptors in the uterus, breast, and other parts of the female body they do not promote tumor growth as the synthetic hormones do. Phytoestrogens can lessen the discomforts of hot flashes, vaginal dryness, headache, dizziness and depression associated with hormonal changes. Black cohosh tea reduces the levels of luteinizing hormone (LH) produced by the pituitary gland. Luteinizing hormones is one of the first active hormones in pregnancy as it regulates the ovaries. By regulating this hormone alone, the symptoms of hot flashes are greatly reduced.

Moringa

Moringa is known for its miraculous cures. Thought since ancient times to cure 100's of diseases, it is being heralded in our day as a cure for malnutrition. Moringa provides 25% more iron than spinach as well as all the essential amino acids and omega fats needed to be well. It is thought to be a woman's herb, because it helps build the blood needed for healthy and safe pregnancies as well as improves lactation and recovery during the postpartum period. Moringa leaves are packed with essential vitamins and minerals including high values of calcium. Calcium is foundational in the regulation of blood pressure and heart disease is one of the leading causes of death for women in America as well hypertensive complications contributes to maternal mortality world-wide. One of the easiest and most economical remedies to decrease hypertension in pregnancy as well thwart preeclampsia is moringa and garlic[xxii].

Moringa and Garlic Tea

1 cup moringa leaves

1 clove of garlic

2 cups water

Bruise moringa leaves and crush garlic.

Pour boiling water over and allow to steep until cool.

Strain, add honey and lemon (optional)

Women should be encouraged to drink this daily and three times daily beginning at 20 weeks in pregnancy.

PMS Herbal Remedy

1 tsp Chaste Berry

1 tsp Black Cohosh

1 tsp Angelica

1 Tbs. Nettles

1 tbs fresh Ginger

Drink daily for a nourishing PMS tonic, but do not drink during the menstrual cycle. Also remember, chaste berry should not be consumed more than 6 weeks at a time. A suggested regiment might be drink the tea for 1 month and take the net month off to all your body to regulate. Or, a more easy tonic might be to make an herbal vinegar tincture and take daily. To make your tincture combine equal parts of each herb and cover with raw apple cider vinegar or alcohol. Allow to steep 6 weeks, shaking daily, then take 10-20 drops in a glass of water daily as indicated above for the tea.

Cramp Relief

1 tsp. Red Raspberry Leaf

1 tsp. Lemon Balm

1 tsp. Cramp bark

1 tsp. Ginger

1 tsp. Black Cohosh

Brew hot. Drink hot or cold all day long.

Cramp Relief Poultice

In a bowl grate 1 large knob of fresh ginger, if possible use a zester. Add boiling water the ginger and make a thick paste. Spread the paste on cheese cloth or a thin cotton cloth and fold. Place over the lower abdomen. Place heating pad on top of poultice and leave for 15-30 minutes or as desired. Massaging the abdomen with evening primrose oil can help to alleviate menstrual cramps

Poultice

A poultice is moistened herbs which have been macerated and placed topically on the skin to aid in the reduction of inflammation or to draw out poisons, infections or toxins.

Popular Poultice Herbs Include:

Comfrey- draws out toxins

Catnip-relieves pain and muscle spasms

Lobelia- relieves pain and muscle spasms

Plaintain- draws out toxins

Add:

Ginger-promotes circulation

Cayenne-promotes circulation

Licorice

Licorice not only helps alleviate coughs, but also is excellent in regulating the hormones cortisal and testosterone and estrogen. Heightened testosterone levels in women can cause polycystic ovarian syndrome which leads to infertility and weight gain[xxiii].

Red Clover

Red clover is considered to be one of the richest sources of isoflavones, which are phytoestrogens. Red Clover is used to alleviate symptoms of hot flashes, PMS symptoms, breast health, improves urine production prevents clotting and bone degeneration.
Red clover is also a source of many valuable nutrients including calcium, chromium, magnesium, niacin, phosphorus, potassium, thiamine, and vitamin C. Bone mineral density is compromised as women age and estrogen production lessens. The use of red clover as a daily tonic have shown remarkable effects on maintaining bone density.

Red clover is also beneficial for women with normal levels of hormone and isoflavones may displace some natural estrogens, possibly preventing or relieving estrogen-related symptoms, such as breast pain by reducing the symptoms of the monthly cycles, and lowers the potential for developing estrogen related cancers.

Red clover can also be used topically and should be considered when making breast or diaper creams.

Violets

Violets are full of Vitamin A and C stores, calcium magnesium and rutin, which is a bioflavanoid. Violets are a mild tonic. The lare stores of Vitamin C and rutin make violets a go to for venous conditions associated with pregnancy and women as they age including: hemorrhoids, spider veins, varicose veins and broken capillaries.

Violets are high in mucilage, which is the sticky substance excreted by the plant when the leaves are bruised of broken. Mucilage will help reduce constipation and in turn reduce hemorrhoids. Violets are also well known to promote breast health.

Lady's Mantle

Lady's mantle works to alleviate symptoms associated with the menstrual cycle. This herb also is used to decrease excessive menstrual bleeding. Lady's mantle can also be used as a uterine prep prior to labor to prevent postpartum hemorrhage. This herb should not be taken prior to 38 weeks in pregnancy.

U(terus) Be Happy Tea

- Lady's Mantle
- Red Raspberry Leaf
- Lemon Balm
- Nettles
- Violets

Yeast Infection

Yeast infections are a given if you are a woman. At some point in your life, specifically in pregnancy, you are likely to have a yeast infection. Some women are more prone to them than others. There are over the counter remedies that are strong, but consider trying a natural remedy first. They can be just as effective and without any added harsh chemicals.

If you are prone to yeast infections consider using *Coconut Oil* as sexual lubrication rather than the store bought lubrication, which again, has all manner of added chemicals and can irritate sensitive vaginal skin. Coconut oil is anti-fungal and can help reduce the growth potential of yeast.

Should you get a yeast infection try fighting it with a garlic suppository. Peel one large clove of garlic and insert vaginally each night for up to 14 days (or for the same length of time as an over the counter or prescription cream.) Each morning bare down over the toilet and release the garlic clove. Gravity should help it to come out easily. If you are concerned, then you can tie a cotton string to the clove and gently extract it in the morning.

Garlic is an anti bacterial. This means it will kill the bad bacteria in your vaginal tract. This also means it will kill the good bacteria in your vagina and now you will need to replace it with good bacterial cultures.

Each morning you will want to use a plain, no sugar added, yogurt that has live cultures to reintroduce good bacteria. You can use a disposable spoon or your clean hands to introduce the yogurt into the vaginal tract. Be sure to wear a thin pad to prevent soaking your underwear.

Rinsing externally with apple cider vinegar is also a great way to reduce the growth potential if the yeast, but if the vaginal tissue is raw this may burn quite a lot. If you are prone to infections consider using this as a prevention, not a cure.

Notes_____
_____
_____
_____
_____
_____
_____
_____
_____

Herbs for Reproduction

Cervical Mucous

Healthy cervical mucous promotes fertility. An over or under production of mucous in the body is not ideal. The following herbs help to support and balance the production of mucous including cervical mucous and helps to balance hormones and restore proper fertility function.

Dandelion
Evening Primrose Oil,
Licorice
Marshmallow
Red Clover
Yarrow

Endometriosis

Endometriosis is the inflammation of the uterine lining and is associated with irregular and extremely painful menstrual cycles.

Ashwagandha: supports both the immune and endocrine systems.

Burdock: breaks down congestion and helps to eliminate toxins through the liver. Supports overall hormonal function and balance.

Cinnamon: improves circulation particular good for the reproductive system

Angelica: improves circulation and helps eliminate toxins.

Echinacea: supports and builds the immune system

Ginger: increases circulation and reduces inflammation

Horsetail: promotes tissue growth and function.

Nettles: supports liver function and balances hormones

Red Raspberry: tones the uterus and helps to reduce heavy bleeding associated with endometriosis

Yarrow: promotes circulation and reduces heavy menstrual flow.

Herbs do not contain progesterone, but they can support the body's natural ability to produce progesterone. Herbs can contain phytoestrogens and estrogen dominance is not healthy either. Understanding and supporting the endocrine system can aid in the balancing of hormones. Adjusting hormones through herbs requires a significant understanding of the body systems and can be quite complicated. The following are herbs that help to promote progesterone production or supports the endocrine system to help balance out both progesterone and estrogen.

Progesterone Support

Alfalfa
Ashwagandha
Burdock
Vitex

Estrogen Support

Burdock

Dandelion root
Evening Primrose Oil
Flax seed
Licorice
Milk Thistle
Red Clover

Amenorrhea (No Flow)

Herbs that help stimulate the uterus and bring on menstruation are known as emmenagogues. A regular menstrual cycle makes conception much easier.

Black Cohosh -Tones the uterus and supports the shedding of the uterine lining.
Angelica- tones the uterus and balances hormones.
Mudwort- stimulates the uterus and promotes menstruation
Parsley- promotes menstruation
Chaste Berry- balances and regulates hormones as well promotes ovulation
Yarrow- stimulates and regulates menstruation

Menorrhagia (Heavy Flow)

When choosing herbs to aid in menstrual flows which are too heavy you will want an herb with an astringent action as well as blood building.

Cinnamon
Ginger
Hibiscus
Nettles
Chaste Tree Berry
Yarrow,

Yellow Dock

Dysmenorrhea (Painful Flow)

Black Cohosh- anti-inflammatory
Black Haw- antispasmodic
Chamomile- Anti-inflammatory and antispasmodic.
Cramp bark- antispacmodic
Angelica- antispacmodic
Ginger- anti-inflammatory
Motherwort- antispasmodic
Red Clover- nourishing and toning
Wild Yam- soothes smooth muscle tissue

Immune Stimulating Herbs for Fertility

Ashwaganda
Angelica
Echinacea
Licorice

Miscarriage

The three herbs commonly used to prevent or reduce the chance of miscarriage are: **black haw, cramp bark and false unicorn.** There is no guarantee a miscarriage, or pre-term labor can be stopped with herbs. The reasons for the contractions must be noted, but the following are remedies tried throughout many cultures with positive results especially where nutritive or hormonal deficiency, stress or tired uterine muscles exists. These herbs are preventative not necessarily curative and should be taken prior to pregnancy in order to establish a healthy uterus. They can be taken at the first sign of miscarriage, but due to the nourishing and slow nature of herbs they should be taken prior. Cramp bark, black haw and false unicorn are effective in reducing uterine contractions and spasms.

NOTE: Never try or recommend these remedies without first consulting with the primary care provider.

Cramp Bark

Cramp bark is most commonly used to inhibit a miscarriage or pre-term labor.

Steep 1 tablespoon cramp bark in boiled water until cool.

Or 20-30 drops of tincture in 1 cup water

Drink

Cramp bark should not be used more than three consecutive days.

Chaste berry and **wild yam** can be used to support the body to inhibit a threatened miscarriage where low progesterone is the cause. They can be combined with cramp bark and black haw.

Polycystic Ovarian Syndrome (PCOS)

This has become an epidemic as of late. Most likely to in large by our modern diets. The herbs used to support a woman with PCOS would need to balance the hormones, as well decrease inflammation.

Ashwagandha
Cinnamon

Burdock
Licorice
Chaste Berry

Herbs for Conception

Choosing herbs prior to pregnancy is all about nourishing the body, balancing the hormones, building the blood and supporting proper uterine and reproductive function. These herbs should not be continued once conception has occurred, so plan to use them 6 months prior to desired conception and discontinue use 4 weeks prior to trying to conceive.

Alfalfa
Ashwagandha
Burdock
Dandelion
Evening Primrose Oil
Hibiscus
Lemon Balm
Nettles
Oatstraw
Red Raspberry

Stress Related Infertility

Chronic stress can cause a multitude of reproduction issues. Prior to conception consider the following for nervine support:

Chamomile
Lemon Balm
Catnip
Passion Flower

Uterine Fibroids

Uterine fibroids are considered to be a condition related to estrogen dominance. Balancing the amount of estrogen one has and supporting the body to naturally produce progesterone will help this condition.

Black Haw: promotes circulation and removes toxins
Cramp bark: soothes irritable uterine cramps and spasms
Ginger: increases circulation and reduces inflammation
Red Raspberry: regulates menstrual blood flow, nourishes and tones the uterus
Chaste Berry: supports hormonal balance and reproduction
Yarrow: promotes uterine circulation and decreases menstrual flow

Uterine Health

Nourishing the uterus is not something we often think of, but we should. The uterus works hard and should be nurtured. The following herbs promote uterine health.

Evening Primrose Oil: nourishes and tones the uterus
Hibiscus: aids iron absorption and nourishes the uterine lining
Nettles: one of the most nourishing herbs for women, it also nourishes the uterus.
Red Clover: aids in purifying the blood and balancing hormones.
Red Raspberry: uteruine tonic

Feminine Steams

Feminine steams work to open up and release toxins and residue from previous cycles. They also work to promote circulations, specifically to the reproductive tissues, which can then promote and regulate menstruation.

Burdock Root: balances hormones

Mugwort: stimulates the uterus and regulates menstruation

Dandelion Root: stimulates digestion and balances hormones

Yellow Dock Root: supports and stimulates the liver to remove toxins

Licorice Root: supports the endocrine system and balances hormones

Ginger: increases circulation and removes toxins

Raspberry Leaf: classing toning herb for the uterus

Rosemary: stimulates blocked menstrual flow

Lavender: cleanses the vaginal tissue, soften perineal scar tissue

Oregano: promotes menstrual flow

Calendula: cleanses the vaginal tissues, softens perineal scar tissue

Sheherd's Purse: Reduces heavy menstrual flow

Basil: promotes menstruation

Aloe: stimulates blocked menstrual flow

Thyme, lady's mantle, and yarrow all work to decrease excessive menstruation flows.

Crampbark, peppermint, false unicorn and catnip relieves painful cramps and bloating associated with premenstrual syndrome and menstruation.

False unicorn: strengthens the ovaries and endometrium and regulates menstrual flow.

Herbs for Menopause

Due to phytoestrogens many herbs can be an incredible alternative to hormone replacement therapies for women either approaching or in the midst of menopause. Most herbs which support the endocrine system will help when going through menopause. When the hormones that course through our blood in our early years taper off, our risk of heart disease rises. The herb hawthorn, used for hundreds of years, has been shown to help conditions like congestive heart failure and high blood pressure.

Angelica- reduces hot flashes
Evening Primrose- estrogen
Black Cohash- reduces hot flashes
Chaste Berry- progesterone and estrogen
Red Clover- estrogen
Hawthorne-cardiovascular health[xxiv]

Herbs for Pregnancy

As a general rule women should not use any herbs in the first trimester. However there are exceptions. Always, always be mindful of your body and how it responds. Discontinue use immediately should you notice uterine cramps.

Herbs that can commonly be used throughout all stages of pregnancy including First Trimester: Echinacea, ginger, chamomile, and red raspberry.

Second Trimester: peppermint and garlic, catnip

Third Trimester: nettles, alfalfa, black haw, false unicorn root, squaw vine, valerian, and evening primrose.

Concerns and Discomforts in Pregnancy

Anemia:

Nettles, Alfalfa, Red Raspberry, Floridex, Hemaplex, Beet Root, Dandelion greens and black strap molasses can all build the blood.

Bladder Infection:

Start with pure cranberry juice with no sugar and echinacea.

Colds

2 hibiscus flowers (without stamens) or 1 Tbs. dried
1 small knob of Echinacea
1 small knob ginger
1 clove garlic
1 cup water

Steep fresh in water and add honey to taste. Drink daily.

Yeast

Insert a clove a garlic during the day and use yogurt with live cultures at night.

Morning sickness:

Ginger
Magnesium

Pregnancy Tea Blend

Nutrition is foundational to healthy pregnancy and birth outcomes. When essential vitamins and minerals are missing from the diet, both mother and baby will be deficient and unable to meet the physiological demands for healthy growth and development. Consuming three cups of this herbal blend daily will increase overall health and well-being for both mother and baby improving outcomes.

Red Raspberry Leaves: Contain vitamins A, B, and E, Calcium, Phosphorous and Iron. This herb is a tonic and helps to promote uterine health and tone.

Moringa Leaves: Moringa contains every essential vitamin and mineral a pregnant woman needs. When dried the protein content is increased providing one of the most deficient nutrients needed to build cells.

Nettle: Nettle is high in iron and Vitamin K. It is a blood builder and helps the body to increase clotting factors. Nettles is also a purifier and supports both the liver and the kidneys during pregnancy.

Alfalfa: Like nettles, alfalfa is high in Vitamin K, Calcium and iron and contains many nutrient vitamins.

Hibiscus or Violet Flowers: Hibiscus is high in vitamin C and can help the body better absorb the irons listed above.

Oatstraw: This herb nourishes the nervous system. Oatstraw is rich in calcium and magnesium. Two essential minerals in pregnancy.

Lemon Balm or **Peppermint** can be added to enhance the flavor of this blend.

Peppermint Oil Vapors

3-5 drops peppermint oil

Epsom salts

A small jar

This aromatherapy can help to ease nausea associated with pregnancy.

Alfalfa

Alfalfa is an incredible source for vitamins and minerals containing: A, D, E, K, and B complex as well biotin, calcium, folic acid, iron, magnesium, potassium and many others. When dried, alfalfa is high in protein and essential cell builder often lacking in diets across the world. Alfalfa has been considered as "the king of foods". Each of these nutrients are beneficial for women suffering from hormonal imbalances, infertility, are pregnant or lactating. Alfalfa actually stimulates breastmilk production. Alfalfa is also high in vitamin K which when taking during the third trimester can decrease the likelihood of postpartum bleeding. Alfalfa blends well with other tonic herbs such as nettle and mints. Oat straw is rich in calcium and magnesium, essential to cell metabolism. This botanical can also help relieve anxiety and restlessness.

Notes _____

_____
_____
_____
_____
_____
_____
_____
_____

Raspberry Bath Tea

2 cup red raspberry leaf

2 cup oats

4 cups epsom salt

4 cups water

Bring water to a boil and add oats and red raspberry leaf to a coffee filter, nylon, or sock and steep. Add Epson salt and dissolve. This bath will help soothe achy or spasmodic muscles muscles and itchiness associated with pregnancy.

Herbs for Labor

Labor Prep

Insert 4-6 capsules of **Evening Primrose Oil** vaginally prior to bed for the last 3 weeks of pregnancy. This will soften the cervix preparing it for the onset of labor.

Labor-Aid Drink

1 Fresh Squeezed Lemon or Lime

1 Fresh Squeezed Orange

½ tsp Sea Salt

2 crushed calcium and magnesium tablets

Honey to taste

4 cups coconut water

There are many variations on this drink and you can make any necessary adjustments in order to suit your flavor palette. This drink is an oral re-hydration therapy and replaces vital minerals and electrolytes, both of which are quickly lost during the exertion of labor. These minerals help the cells to function properly and are essential to the coordination of muscle fibers and function which is so very important in labor and childbirth.

Valerian is wonderful to use in tincture form to help a woman relax in labor. Combine it with skullcap to take the edge off if a mom feels a little out of control. It can also be used topically in massage oils to relax tense muscles.

Trillium (Bethroot) is used in early labor to smooth the progress of contractions and transition a mother from early into active labor. This herb supports uterine contractions and can make for an easy labor and birth. It is also thought to decrease the onset of postpartum hemorrhage.

Cramp bark can be used in transition to soothe the uterus. It is an antispasmodic and will help the uterus relax and use the energy for deep contractions which melt away the last of the cervix. Cramp bark is considered safe to use in all stages of pregnancy and is thought to help prevent postpartum hemorrhage when used in labor and is also useful in postpartum to help ease after pains.

Squaw Vine is a uterine tonic and can be used in the last weeks of pregnancy to help transition the uterus into a good consistent rhythm of contractions. Can steep and drink as a tea daily for the last 4 weeks of pregnancy.

Catnip Leaf is a mild sedative and can ease restlessness and birth anxiety. It can also be used in the last trimester to help promote sleep. Steep in boiling water. Add honey, lemon and ice and sip cold.

Passionflower can be added to this tea when a mom is in labor to ease anxiety.

Herbs for Postpartum

Postpartum Tea

1 cup hibiscus flowers

½ cup chamomile flowers

½ cup oat straw

½ cup red raspberry leaf

¼ cup lemon balm

¼ cup lavender flowers

Add honey to taste

Add to a gallon of water and bring to a boil. Turn off and allow to steep until cool. Strain and drink until gone. Store in a glass container in the refrigerator.

Notes_____

_____
_____
_____
_____
_____

Afterbirth Tincture

Catnip

Cramp bark

Black Haw

Skullcap

Add equal parts of the above listed herbs to a pint or quart jar. Add vodka until the herbs are covered and allow to steep for 6 weeks. Shake daily and add more liquid as necessary to keep herbs covered. Strain and use as needed to ease uterine cramps.

Bottom Bliss Spray

1 cup Witch Hazel

1 cup water

2 large comfrey leaves

3 large aloe vera stalks

10 drops lavender essential oil

Chop comfrey leaves and infuse in water. Bring water to a boil then add chopped leaves and steep until completely cool

Open the aloe stalks and scrape out the gel.

Strain the comfrey and add the aloe gel, witch hazel and essential oil. Pour into a glass spray bottle or a peri bottle and use over the perineum.

Ashwaganda and Gotu Kola Tincture

Ashwaganda and Gotu Kola tincture can be taken daily to support the nervous system. Ashwaganda is known as an adaptogen and can help the body cope with the stress of the postpartum period. Gotu kola is a nervine and also soothes the nerves.

1 pint jar

Fill 50/50 with ashwaganda and gotu kola

Cover with apple cider vinegar or vodka and allow to steep 6 weeks. Shake daily and add more liquid as necessary to keep herbs covered. Strain and take 2-3 droppers full in a glass of water to soothe nervousness associated with postpartum.

Postpartum Sleep Tonic Tincture

Hops

Skullcap

Lavender

Valerian

Add equal parts into a quart jar and cover with 80 proof brandy or vodka. Shake daily and add more liquid as necessary to keep herbs covered. Allow to steep for 6 weeks. Strain and store in a glass jar.

Take 2-3 droppers full in a glass of water prior to bed or during the day after baby is down for a nap. Or take 1-2 dropper full in a glass of water to ease anxiety.

Lemon Balm Tea

This herb is a gentle nerving and renown for lifting the spirit. Use in postpartum to stave off the baby blues.

3 Tbs lemon balm

¼ teaspoon lavender flowers

Bring 1 cup of water to boil and pour over the herb sachet. Add honey to taste and sip when warm.

Herbal Bathe

1 cup Comfrey Leaf

½ cup Lavender Flower

1 cup gotu kola leaves

1 cup Plantain Leaf

½ cup Red Raspberry Leaf

¼ cup Yarrow Flower

¼ cup Calendula Flowers

¼ cup Shepherd's Purse

1 Bulb garlic

½ Sea Salt

½ Epsom

Mix together the dried herbs and salts.

Draw a hot bath, add 2-3 cups of herbs tied together in a sachet and allow the bath to cool until warm. Soak in tub.

Herbs for Breastfeeding

Herbs recommended for increasing milk supply and promoting lactation are know as galactoglogues. These are herbs which stimulate milk production and often are ones which also aid in digestion or are overall tonic in nature full of many nutrients needed for proper healing and recovery in the postpartum period. Any of the following can be used to support the postpartum period and to promote lactation and can be taken as tinctures, teas, infusion or in a combination.

...better as tea

Fennel

Fenugreek

Anise Seed

Blessed thistle

Red Raspberry Leaf

...better as tincture

Hops

Chaste Berry

Goat's rue

Milk Thistle

...better as infusion

Nettle

Dandelion

Nipple Cream

4 Tbs bee's wax
2 Tbs. infused olive oil (infuse with plantain and/or comfrey)
3 Tbs. coconut oil
3 drops calendula essential oil

3 drops lavender essential oil
Melt the bee's wax and olive oil together over low heat. Add in the coconut oil and essential oils. Place in shallow container and allow to cool. * Add more wax for a firmer cream.

Itchy Nipples (Yeasty Beasties)

2 Tbs apple cider vinegar

1 cup water

3-5 drops tea tree oil

Mix and spray over nipples after feeding then pat dry.

Echinacea and Garlic Tincture

Fill a pint jar with chopped Echinacea root and raw whole garlic cloves 50/50.

Cover with apple cider vinegar and allow to infuse for 6 weeks. Shake daily and add more liquid as necessary to keep herbs covered. Strain and use as an antibiotic at first sign of mastitis.

Mastitis Poultice

1 Tbs Comfrey

1 Tbs Calendula

1 Tbs Rosemary

1 Tbs Garlic

1 Tbs Dandelion

4 cups water

Bring water to a boil and add herbs. Turn down and allow to simmer until water is half evaporated and the infusion strong. Turn off and allow to cool. Strain herbs, wrap in cheese cloth or muslin and use as poultice. Keep water, store in glass container in refrigerator and use as a rinse 3 times daily until gone.

Herbs Contraindicated During Lactation

Aloe Vera gel

Bugle-weed

Butterbur

Coffee beans

Comfrey (internally)

Jasmine flowers

Kava-kava root

Herbe mate

Oregano

Parsley

Peppermint

Sage

Sorrel

Thyme

Notes_____
_____

Herbs for Newborn

Herbs are not recommend for babies internally until at least 6 months of age, but can readily be used topically with little concern. Let us consider however, if this recommendation is accurate. If herbs, as mild as these, are not recommended for infant use then why should we be willing to give harsh chemical laden immunizations to that same infant? Be wise and knowledgeable. Know which herb you are using and why, choose organic and only mild herbs such as chamomile, lavender, dill, fennel, catnip or lemon balm and always use the correct dose.

Soothing Newborn Bathe

½ cup lavender flowers
½ cup calendula flowers
½ cup chamomile flowers
1 cup plaintain
1 cup comfrey
1 cup sea salt

These herb relax, sooth and help promote healing. They are anti-inflammatory, antimicrobial and antiseptic in nature. Baby should not be bathed in the first 48 hours of life to help them maintain anti-bacterial microbes naturally occurring on their skin.

Add 1 cup to a hot bathe and allow to cool until just warm. Bathe baby.

Washing Powders

6 cups washing soda

1 Tbs. Baking soda

2 bars castille soap

20 drops lavender essential oil

20 drops orange essential oil

Grate the bars of soap and add with washing powders. Add essential oils and shake to mix. Store in a glass jar and use 2 Tbs. with each load.

Diaper Salve

4 Tbs. grated bee's wax
2 Tbs. olive oil
2 Tbs. coconut oil
2 drops myrrh essential oil
2 drops lavender essential oil
2 drops frankincense essential oil

Warm the oil in a pan on the stove and slowly stir in the wax until melted.

Cradle Cap

1 Tbs crushed fenegreek seeds

1 tsp. Olive or Coconut Oil

Make paste and apply to scalp.

Allow to sit for 15-20 minutes.

Using warm water and a soft bristle brush gently loosen the scales.

Rinse with warm water.

Sleep Water

1 tsp lavender flowers

1 tsp chamomile flowers

1 cup water

Bring water to boil and add flowers. Steep until cool. Give 1 tsp. as needed for sleep. Store in a glass container for 24 hours in refrigerator.

Warm Castor Oil Rub

This rub helps to relieve gas in an infant and can help alleviate colic.

2 drops lavender essential oil

2 Tbs. Castor oil

Warm oil in a pan on the stove. Remove from heat and allow to cool. Add essential oil at this time. Dip fingers in the warm, but cooled oil and begin at the belly button and rub in a clockwise direction massaging very lightly and gently. Rub for 5-10 minutes and apply more oil as needed.

Gripe Water

1 tsp. Chamomile

1 tps. Lemon Balm

1 tsp. Catnip

1 tsp. Fennel Seed

1 tsp. Dill seed

1 cup water

Wrap herbs in a coffee filter and tie with a string. Bring water to boil and add homemade sachet. Turn off and allow to steep until completely cool. Store in a glass container for up to 24 hours. Give a colicky baby 10 drops.

Newborn Massage

1 cup organic olive or coconut oil

¼ cup chamomile flowers

3 drops lavender essential oil

Warm oil in a pan on the stove and gently infuse organic olive or coconut oil with chamomile flowers. Cool and add essential oils. Store in a glass container.

Notes_____

_____
_____
_____
_____
_____
_____
_____
_____
_____
_____
_____
_____
_____
_____
_____
_____
_____
_____
_____
_____

"If you can't pronounce the word, don't eat it and don't rub it all over your body."

Amy Kirbow

Herbs de Toilette

Using herbs as an alternative to common cosmetics, and basic grooming products is advisable for many reasons. Creating your own product allows you to determine what will and what will not be in the product and harsh undesirable chemicals are completely excluded. You are able to maintain quality control and only use the best ingredients, preferably ones you or your community has grown. Years ago, I began to experiment with making my own face wash, shampoo, night cream , toothpaste along with many other products. Some I honestly found to be of better quality to purchase from reputable sources, but others were not only easy to make, but less costly and of equal to greater value

Oatmeal Honey Scrub

1 cup oats ground up finely in a blender
2 Tbs honey (or more to form firm paste)
1 tsp. Nutmeg
1 tsp. cinnamon
1 Tbs. coconut or olive oil

Mix together to form paste. Store in a glass container.

Facial Steam

Calendula
Chamomile
Lavender flowers
Orange or other citrus peel

Peppermint

Rosemary

Rose petals

Add ½ cup of any of the above listed herbs to a pan of water. Bring to a boil. Hold face over the steam and cover head and shoulders with towel to prevent steam from excapein. Use caution and do not burn yourself.

Facial Mask

2 Tbsp plain yogurt

1 egg white or yolk

5 drops lavender essential oil

1 Tbs. honey

5 drops coconut or olive oil

Mix together and apply to face. Store for 24 hours in a glass container.

Hand Salve

2 cup coconut oil

1 oz grated bee's wax

10 drops lavender essential oil

Add oil and wax and slowly melt in a pan on the stove. Add more wax for firmer salve and add essential oils. Store in a glass container.

Lemon Hand Scrub

1 cup sugar

2 Tbs olive oil

10 drops lemon essential oil

Mix and use. Store in a glass container.

Coffee Facial or Body Scrub

1 cup coffee grounds
1 cup sugar
1/2 cup coconut oil (whipped)
1 Tbs. cinnamon
1 Tbs. nutmeg
3 Tbs honey

Mix and use. Store in a glass container.

Body Scrub

¼ cup Epsom salt
1 Tbs. olive oil
5 drops lemon essential oil
5 drops lavender essential oil
Mix and use. Store in a glass container.

Moisturizing Body Scrub

1 cup sugar

1 cup coconut oil

Whip the (firm) coconut oil in a stand mixer fitted with the whisk attachment. Whip until fluffy. Add the sugar and whip until combined.

Turmeric Scrub

.1 tsp turmeric powder

2 tbs honey

½ cup ground oats

Add to blender and blend to make paste. Add more honey as necessary.

Mouth Wash

1 cup water

1 teaspoon whole cloves

1-2 cinnamon sticks

10 drops myrrh essential oil

Bring the water to a boil and add cloves, cinnamon and peppermint and remove from heat. Let the spices infuse until cool.

Strain the mixture and store in a glass container. Refrigerate.

Shampoo

1/4 cup coconut milk

1/4 cup castille soap

5 drops coconut oil

1 tsp baking soda

10 drops rosemary essential oil

10 drops lavender essential oil

Combine. Shake to mix and use. Stores for about a month.

Toothpaste

1/2 cup coconut oil

2-3 Tablespoons of baking soda

5 drops of peppermint essential oil

5 drops lavender essential oil

5 drops myrhh essential oil

Stevia can be added to sweeten.

Combine and store in a glass container.

Notes

Herbs for the Household

There are so many ways herbs can be used around the house. Learning to make your own household products is the ultimate way to incorporate herbs into your every day life on a whole new level and to remove just one more level of chemical exposure. Often, we do not think about the chemicals we expose ourselves to daily through detergents, cleaners and soaps. Returning to the basic will not only be economically fulfilling, but extremely beneficial to our health and the health of our family.

Rosemary, Thyme and Basil Counter Spray

5 drops basil herb essential oil
5 drops thyme herb essential oil
5 drops rosemary herb essential oil
1 cup white distilled vinegar
½ cup water

Mix all of the above and add to a spray bottle.

Tub Scrub

¼ cup baking soda
1 tsp. castile soap
½ cup water
5 drops orange essential oil
5 drops lemon essential oil

Mix and use.

Toilet and Bathroom Spray Cleaner

1 cup distilled white vinegar

½ cups water

1 Tablespoons castile soap

10 drops lavender essential oil or

10 drops peppermint essential oil

Orange Infused All Purpose Cleaner

2 cups white distilled vinegar

1 cup orange peels

10 drops orange essential oil

Add peels to a glass container and pour the vinegar over the peels. The liquid should cover the peels, if not, add more. Cover the container and allow to mellow for about 2 weeks in a dark place. Strain, add essential oils and use.

All Purpose Cleaner

1 cup of water

½ cup white distilled vinegar

10 drops lavender essential oil

Mix, add to a spray bottle and use.

"A weed is a plant whose virtue is not yet known."

Ralph Waldo Emerson

Foraging Herbs

Learning to forage means you are never without a pharmacy at your finger tips. Once you know how to forage for herbs your medicine cabinet will be self-sustainable and always full. There are three rules in foraging: 1. Identity 2. Identity and 3. Identity. You absolutely must know what you are collecting. Plants with similar features can be deadly. There are enough stories of mistaken plant identity to convince any forager to be vigilant and sure before consuming. When learning to forage begin with plants which are easily identifiable and with few mistaken counterparts such as dandelion, plantain, violets or mullein. When foraging, it is best to go early in the morning before the sun has become too strong. Always choose the healthiest plants and never take all of any one plant from an area, especially if you are taking the root of the plant. There are many reasons foraging should be learned, not only is it free, but you are also forced to be out in the sunlight (vitamin D) and exercising at the same time. If you will being to forage you will find yourself in the garden relationship we began this book with. You will be walking in nature, and talking to God, should you choose, all while learning how to care for your self, mind body and spirit.

Notes_____
_____
_____
_____
_____
_____
_____
_____
_____
_____

"Of course it works... I made it myself out of dried weeds and vodka!"

Jim McDonald

Recipes From My Herb Cabinet

Warming Winter Tea

4 cups water

1 lemon

1 orange

4 cinnamon sticks

1 knob of ginger

1 tsp. Cloves

4 black tea bags

4 cups water

Bring water to boil and steep above ingredients. Add honey to taste

Calcium Building Vinegar

¼ cup Dandelion

¼ cup Nettle

¼ cup Parsley

¼ cup Plantain

¼ cup Violet

Add to large class jar and cover with raw apple cider vinegar. Allow to mellow for 6 weeks, shaking daily. Refrigerate and take daily for a calcium and mineral boost.

Tummy Tea

1 Tbs. peppermint

1 knob of ginger

1 Tsp. Fennel seeds

1 Tbs. lemon balm

Add to 1 cup boiling water and allow to steep. Add lemon and honey to taste.

Easy Me Tea

2 Tbs. chamomile

2 Tbs. lemon balm

2 Tbs. catnip

½ tsp lavender

Steep in pint of water, Add honey prior to cooling and dissolve. Refrigerate and drink cold.

Whipped Coconut Oil

2 cups (firm) coconut oil
10-15 drops essential oils

Place oil in a stand mixer and use the whisk attachment. Whip until sufficiently fluffy. Add essential oils and whip just until blended.

Feminine Blend Tea

1 Tbs. Alfalfa

1 Tbs Red Clover

1 Tbs Nettle

1 Tbs Yarrow

1 Tbs Oatstraw

1 Tbs Fennel Seed

1 Tbs Lemon Balm

1 Tbs Hibiscus Flowers

1 Tbs Peppermint

1 Tbs Chaste Berry

Steep in 1-2 quarts of boiling water. Add honey to taste. Allow to cool. Drink warm or refrigerate and drink cold.

Magnesium Spray

1 cup magnesium flakes
1 cup water
5 drops peppermint essential oil

Add 1 cup of magnesium flakes to 1 cup of boiling water and stir until dissolved. Place in a spray bottle and apply to skin daily. Add 5 drops of lavender essential oil.

Nourishing Vitamin and Mineral Formula

1 Tbs. Garlic

1 Tbs. Alfalfa

1 Tbs. Oat straw

1 Tbs. Nettles

1 Tbs. Red Raspberry Leaf

1 Tbs. Parsley

1 Tbs. Comfrey

1 Tbs. Basil

1 Tbs. Cayenne

1 Tbs. Oregano

1 Tbs. Dandelion leaf and/or root

1 Tbs. Thyme

1 Tbs. honey

Add to ½ gallon of water. Bring to a boil, turn off and allow to steep until cool. Drink at room temperature or refrigerate and drink cold.

Hibiscus Tea

Pick 10 red hibiscus blossoms.

Remove stamens and stems and add to 1 quart of water. Bring to a boil, turn off and allow to steep. Squeeze in half a lime in order to activate the red color. Add honey to taste. Sip warm or drink cold.

Also tasty add any of the following: cinnamon sticks, orange, extra lime or lemon, and peppermint.

Immune Building Tea

2 Tbs fresh ginger grated

2 Tbs. fresh turmeric grated

1 Lemon squeezed

1 cup boiling water

Allow to steep. Add honey and milk to taste.

Nettle and Mint Infestation

1 Tbs nettles

1 Tbs. peppermint

1 cup boiling water

Add honey to taste

Allow to cool and drink over ice.

Herbal Reference

Alfalfa	Blood cleanser, aids in digestion and builds immunity. Alfalfa is a source of many vitamins and minerals including vitamin K which helps clotting.
Aloe Vera	Mainly used topically as a skin tonic, but can be taken internally to improve digestion. Used commonly for sunburns, burns, scrapes, and to prevent infection in minor wounds. Ingesting too much can have laxative effects and leads to cramping and diarrhea.
Angelica	Known traditionally as a birthing herb, angelica is used to induce labor and bring a retained placenta. It is also used to relieve painful cramping associated with menstruation.
Anise Seed	Aids digestion, alleviates gas and nausea, can be used to help reduce the symptoms of morning sickness.
Black Cohash	Like angelica, black cohash is used to alleviate menstrual irregularities and bring on menstruation. It can also be used to induce labor. It is often used during menopause to reduce hot flashes, vaginal dryness and associated depression.
Black Haw	Used to alleviate menstrual related uterine spasms this herb can also be used to prevent miscarriage and afterbirth pains. Black haw is also a diuretic and can be used to increase urine output.
Blue Cohosh	Like black cohosh, blue cohosh is used in a similar manner. Often they are used in tandem when inducing labor. Blue cohosh stimulates the uterus and helps to bring on menstruation.
Boneset	This herb is used to induce sweating and can help treat colds and flues, as well as fever and symptoms associated dengue, cholera and malaria.
Calendula	This herb is an antiseptic and is used topically for minor wounds, cuts scrapes, bruises and to prevent infection.
Catnip	One of the most gentle herbs catnip can be given to very young children. Aids in digestion, relieves gas, decreases restlessness and insomnia and is considered a mild sedative.

Cayenne	Increases blood flow whether used internally or externally. Relieves sore muscles when used topically. Also topically can reduce bleeding and cause cauterization. Often used to relieve stiff arthritic joints. Not suggested for use after birth as it can cause excessive bleeding. Cayenne is also considered a wonderful heart tonic.
Chamomile	This herb is an overall tonic and is wonderful for hair and skin. Also, used as a mild sedative, especially when paired with catnip. Considered a stomach tonic as it alleviates indigestion and gas. Topically chamomile is used to wash wounds and can be used as an external wash for vaginosis and yeast.
Cinnamon	An antibacterial cinnamon is used to relieve colds, sore throat and flues. Also can be used alongside chamomile as a rinse for yeast infection.
Coconut Oil	An anti-fungal coconut oil can be used to treat yeast infections and can also be used as a vaginal lubricant. *Caution when using as a vaginal lubricant alongside condoms as the oil may break down the rubber.
Comfrey	Known for is healing properties it has been used for fractures, bruises, open wounds, cuts, scrapes. Also known to relieve pain due to inflammation. Often used in postpartum rinses.
Cotton Root Bark	Used to regulate menstrual dysfunction, to induce labor and to expel the placenta. Cotton root bark is also used to increase milk supply during lactation.
Crampbark	This herb is used primarily to reduce uterine spasms. Can be used during the menstrual cycle or to ease afterbirth pains. Like black haw it can be used when miscarriage threatens.
Cumin	Relieves gas, aids digestion and soothes irritable stomach.
Dandelion	Stimulates digestion and cleanses the liver. Dandelion is an overall tonic, supports kidney function and can be used as an acne wash by reducing excess skin bacteria.
Echinacea	Is immune building and used to treat colds and flues. Echinacea also helps prevent the bacteria which causes yeast infections.

Elderberry	Can be used topically to treat wounds. Also used as a treatment for colds and flues by the reduction of inflammation in the mucous membranes as well as the associated respiratory ills.
False Unicorn	Used as an overall well woman tonic, false unicorn is also used in the prevention of miscarriage. This is herb is also used to treat ovarian cysts, regulate menstrual and menopausal symptoms and by helping to balance hormones.
Fenugreek	Most commonly this herb is used to promote lactation. It is also used to soothe upset stomach and associated constipation. Fenugreek has estrogen properties and can be used to lessen the symptoms of menopause, reduce hot flashes and increase libido.
Feverfew	Most commonly used to treat headaches and is often combined with lemon balm.
Ginger	Ginger is best used to treat stomach ills including: morning sickness, nausea, colic, gas, diarrhea, motion sickness and overall indigestion. Ginger is a warming herb and can be used as a poultice to reduce uterine cramps associated with menstruation or afterbirth.
Goat Rue	Most commonly used to treat diabetes by decreasing blood sugars, goat rue also is used to increase milk production and can also be used to regulate menstruation.
Gotu Kola	Gotu Kola is commonly used in the treatment of bacterial and viral infections including UTI's. It is often used to treat anxiety and depression and to increase memory. Gotu Kola can also be used to treat venous insufficiency leading to varicose veins and is used topically to treat wounds.
Hawthorn	Most commonly used for heart health. It can also be used to aide digestion and decrease anxiety.
Hops	This herb is used as a natural pain remedy as well decreases anxiety, restlessness and associated insomnia. Hops is also very commonly used to increase or bring on milk flow.
Lavender	Lavender is used to aid digestion as well to decrease gas and nausea. It is also is used to decrease anxiety and depression.

© GoMidwife 2015

	Lavender is also used to decrease nervousness and support sleep. Lavender can also be used to promote menstruation.
Lemongrass	Most commonly used for intestinal upset lemongrass soothes stomachaches and digestive tract spasms. It is also used topically in natural bug sprays and to kill germs.
Licorice Root	Most commonly used to treat colic, stomach upset, reduce heartburn and decrease inflammation in the stomach lining. It is also used to treat sore throat and cough.
Mother Wort	This herb is used as a heart tonic. Mother Wort is also used to increase fertility and to bring on menstruation.
Mullein	Excellent for respiratory ills including cough and whooping cough, bronchitis and pneumonia. Mullein is also an incredible skin tonic and is used to treat wounds and hemorrhoids.
Nettle	Nettles is an excellent overall woman's tonic, and is especially good as a urinary, liver and kidney tonic. It is a blood builder and can help to decrease anemia.
Oat Straw	This herb is a nervine decreasing stress and anxiety by soothing the overall function of the nervous system. This herb is high in calcium and micro-nutrients.
Oregano	This herb is anti bacterial and anti fungal and can be used to treat and or prevent yeast and vaginal infections as well is used in the treatment of colds and flues. Oregano also is used to treat acne and dandruff by reducing the bacteria on the skin and can be used to reduce cramping associated with menstrual cramps when applied topically.
Passion Flower	Warning: Passionflower should never be used when also using an MAO Inhibitor. This herb is used to decrease anxiety and nervousness, as well as a sedative to promote sleep especially in combination with skullcap, lemon balm and Valerian.
Peppermint	Topically peppermint is used to relieve headaches. It is notable for increasing digestion and reducing symptoms associated with stomach upset including nausea and vomiting as well as morning sickness. Peppermint relaxes the digestive tract and can

	exacerbate heartburn. This herb is also used to reduce inflammation associated with colds.
Red Raspberry	This herb is a uterine tonic, helps prevent miscarriage, reduces morning sickness, reduces pain associated with heavy menstrual periods and prepares the uterus for labor. It is also used to treat gastrointestinal tract disorders increase the production of bile and urine.
Rosemary	This herb is used to relieve gas, increase and aid in digestion, reduce heartburn and promote liver and gallbladder function. Rosemary is also used to relieve headaches. Topically rosemary aides in wound care, skin care, reduces dandruff, and reduces inflammation associated with joint and sciatic pain.
Sage	Sage is used to correct digestions problems including: gas, heartburn, bloating, diarrhea, and stomachache. Sage reduces the over-production of sweat and saliva as well as the production of milk. Sage is used after miscarriage or still birth to stop lactation. Sage also can be used to regulate menstruation and to wash the symptoms of menopause. This herb also makes and excellent mouthwash.
Shepherd's Purse	This herb is used to ease symptoms associated with PMS, to regulate menstruation and to decrease the length of long menstrual cycles. Topically shepherd's purse can be used to stem bleeding. Often used during postpartum to reduce excessive bleeding by stimulating the uterus increasing contractions it should be used with caution as it can lead to large clots.
Skull Cap	Most commonly used to decrease anxiety and associated nervousness it is also used to promote sleep. When combined with hops skullcap is a mind sedative and helps relieve pain.
Slippery Elm	This herb soothes the gastrointestinal tract and relieves associated constipation, diarrhea, IBS, inflammation and reduces stomach acid. Slippery elm is also used for coughs and colds.
Squaw Vine	Squaw vine is used to treat fibrocystic breasts, regulating menstruation and vaginal discharge. This herb can also be used to

	ease labor and depression associated the postpartum period as well promotes lactation. Squaw vine has also been used to reduce anxiety and regulate the bowels.
Thyme	Thyme is commonly used for gastrointestinal and urinary regulation as well in the treatment of colds, flues and associated symptoms. Thyme is used for treating sore throat and as an mucous expectorant in wet coughs.
Turmeric	This herb is an anti-inflammatory and immune builder. Turmeric is also used to aid digestion, reduce symptoms of colitis and soothe stomach ulcers. It stimulates the gall bladder and the production of bile and can reduce bloating and gas.
Uva Ursi	This herb mainly supports the urinary and kidney system. Uva ursi reduces inflammation of the urethra and urinary tract and can prevent and treat UTI's.
Valerian	Valerian can be used to reduce anxiety, but is most commonly used to promote sleep and reduce insomnia. Often paired with hops, skull cap, passion flower or lemon balm it is considered a sedative.
Vitex	This is a woman's herbal tonic and is widely used to regulate hormones and reduce symptoms associated with PMS. Vitex (or chaste berry) is used to regulate ammenorrhea and dysmenorrhea. It can be used to ease promote and ease labor and to reduce bleeding following birth and to help expel the placenta.
Wild Yam	This herb is used to regulate menopause and the associated symptoms as well as those related to PMS.
Yarrow	Yarrow bathe soothes menstrual cramps and hemorrhoids. This herb can also be used topically for wounds. Yarrow also promotes menstruation.

The best thing we can do for our fellow man is to wake things up that are in them... to make him think for himself.

George MacDonald

Please Note: not all the above mentioned herbs should be taken in combination with one another. Try simplifying by using one herb, tonic or tincture at a time for several days or weeks to see how your body responds. Always be aware of your body and its responses to herbs. Always, always know what herb or botanical you are actually putting in your body, where it was grown, how it was raised and harvested. Use caution when consuming herbs. None of the herbs or recipes in this book have been approved by the FDA and this is both my understanding and opinion on herbs, combined with years of research and experience. Always check with your primary care provider before consuming herbs medicinally and know that you alone are responsible for what you put into your body.

End Notes

i Herbs date back to/Introduction to herbs

http://www.melwood.org/en_us/blog/an-introduction-to-herbs-2

see also

"before the establishment"/intro to herbs

http://exhibits.hsl.virginia.edu/herbs/brief-history/

ii Definition of medicine- www.dictionary.com

iii Counterthink cartoon- chinese herbs

http://www.naturalnews.com/021964_biopiracy_medicinal_herbs.html

iv Heal Yourself

Read more: http://undergroundhealthreporter.com/completely-rebuild-your-body-from-the-inside-out/#ixzz3gxKLz13i

v History of Herbs

https://www.planetherbs.com/history/the-herbal-tradition.html

vi the history of herbs

https://www.planetherbs.com/history/the-herbal-tradition.html

http://www.naturalhealthschool.com/history-of-herbalism.html

vii defining herbs/herbal reference

http://www.herbaleducation.net/herbs-glance

viii definition of pharmakia strongs condcordance

http://www.biblestudytools.com/concordances/strongs-exhaustive-concordance/

ix Counter think cartoon- lets drug half the population

http://www.google.com/imgres?imgurl=http://www.nutritional-supplements-health-guide.com/images/cartoons/ct_birth_big_pharma_600.jpg&imgrefurl=http://www.nutritional-supplements-health-guide.com/what-is-niacin.html&h=600&w=579&tbnid=TjMtm9PuUUfgAM:&docid=uRJmLm8PEY6gvM&ei=0KOlVbeXIMbmoATwirTgDg&tbm=isch&ved=0CC0QMygQMBBqFQoTCLe_8rbr28YCFUYziAodcAUN7A

x 70% of Americans are on...

End Notes

http://www.sciencedaily.com/releases/2013/06/130619132352.htm

xi http://armidwife.com/Herbs_and_Supplements.html

xii http://avivaromm.com/

xiii http://www.medicalnewstoday.com/articles/150999.php

xiv Nourishing herbs-

http://www.prevention.com/mind-body/natural-remedies/25-healing-herbs-you-can-use-every-day/thyme

xv Muscular skeletal system herbs http://www.portlandherbalschool.com/wpcontent/uploads/2010/05/The_Muscular_Skeletal_System.pdf

xvi http://muktiorganicskincare.com/herbs-for-skin/

xvii Lymphatic system herbs

http://www.livestrong.com/article/395149-what-herbs-deal-with-the-lymphatic-system/

xviii Eggshell Tincture

http://www.thehealthyhomeeconomist.com/simple-eggshell-tincture-for-acid-reflux/

xix http://naturallynourishing.com/the-benefits-of-hibiscus-flower-tea/

xx Most info on herbs came from: University of Maryland Medical Centerhttp://umm.edu/health/medical/altmed/herb/

xxi http://natural-fertility-info.com/fertility-herbs

xxii http://imaginezambia.org/projects/moringa-project/moringa-the-miracle-tree/

xxiii Lists of herbs for pregnancy etc.
http://natural-fertility-info.com/fertility-herbs

xxiv Hawthorne
http://jaktraks.hubpages.com/hub/herbs-cardiovascular-system

www.ingramcontent.com/pod-product-compliance
Lightning Source LLC
Chambersburg PA
CBHW080918170526
45158CB00008B/2154